D0872287

Practical Techniques for Laboratory Analysis

James A. Poppiti

Library of Congress Cataloging-in-Publication Data

Catalog record is available from the Library of Congress

Direct all inquiries to CRC Press, Inc., 2000 Corporate Blvd., N.W., Boca Raton, Florida 33431.

Lewis Publishers is an imprint of CRC Press

International Standard Book Number 0-87371-361-3
Printed in the United States of America 1 2 3 4 5 6 7 8 9 0
Printed on acid-free paper

ACKNOWLEDGMENTS

I would like to thank the many laboratory managers I have spoken to and had the opportunity to discuss laboratory operations. You have all had input into this book and for that I thank you. I would also like to thank J. J. McCown for his interest in this project and discussions relating to analytical operations within the Department of Energy. I want to thank John Reed for his help preparing this manuscript. I would especially like to thank my wife Vicky for her support and encouragement that made this book possible.

PURPOSE OF THE BOOK

I have written this book with a single purpose - to provide people who work in laboratories with information they won't find anywhere else. My goal is to provide practical information that you can put into practice right now to make your laboratory run smoother and more profitably. Things that will help you do your job better, whether your job is sample preparation, instrumental analysis, or lab management.

There are a great many things that a person needs to know to work in a modern laboratory that aren't written down anywhere. There is so much information left out of analysis methods, analytical manuals, analytical text books, quality assurance manuals, and management courses that I decided to share my experience in the hope of shortening your learning process. This is why I wrote the book.

The material for this book is the result of my own experience and the combined experience of many veteran lab managers and practicing analytical chemists. In preparing the material for this book I have interviewed laboratory managers and practicing chemists from over 50 laboratories, both large and small, to find "what works" and what does not.

I also present the reasons why you have to do something a particular way, (i.e., "know why" vs. "know how") because this understanding is the basis for real knowledge. I present my experience from learning "the hard way" so you won't have to.

OUR INTENDED AUDIENCE

It is my hope that both novice and experienced professionals find this book useful in reviewing basic and advanced techniques. Chapter 1 of this book is intended more for the laboratory manager. Here I deal with questions of laboratory organization, structure, costs, etc. The remainder of the book is intended more for the professionals and technicians that actually work in the laboratory. In Chapters 2 and 3 I provide details about how analyses are done, why certain steps are important, and how to get more out of your time in the lab.

HOW TO USE THIS BOOK

This book is organized like a handbook. You don't need to read the chapters that don't apply to you. If you are involved with managing the laboratory you should read Chapter 1 and be somewhat familiar with the material in the rest of the book. If you prepare samples for organic analysis you should read Chapter 3.

It is my intention that this book serve as a reference for people working in laboratories. The book provides instruction for most routine laboratory analyses. The analyst, after mastering one method or technique, can read another section of the book and begin working on other analysis methods.

I hope that this book helps you in your everyday work and provides the information you need.

THE AUTHOR

Dr. James A. Poppiti graduated from the University of Delaware in Chemistry in 1971 where he developed his interest in analytical chemistry in general and mass spectrometry specifically. (Burnaby Munson, co-founder of Chemical Ionization Mass Spectrometry, was his advisor at Delaware and hence, his interest in mass spectrometry.) He went to graduate school at the University of Virginia for a few years and eventually took a job with Finnigan Corp. as a service engineer. While at Finnigan the first commercial simultaneous positive/negative chemical ionization controller was developed.

From Finnigan, Dr. Poppiti went to the Food and Drug Administration in Washington D.C., where he analyzed foods for commercial chemicals, including PCBs, pesticides, dioxins, etc. While at the FDA, he was recruited by the Environmental Protection Agency to join a regulatory development group that was writing regulations for the Resource Conservation and Recovery Act (RCRA). At the EPA he was the editor for the second edition of the testing manual "Test Methods for Evaluation Solid Waste" (also known as SW-846).

Dr. Poppiti went back to work for Finnigan for 3 years in the marketing and engineering departments. Most of the material for this book was collected during this time when he visited about 100 commercial laboratories and interviewed their laboratory directors, bench chemists, and technicians about their jobs and the laboratory business. He joined the Department of Energy in 1992 and is currently working in the Environmental Restoration and Waste Management Program.

TABLE OF CONTENTS

Chapter 1
Laboratory Operations — Design and Management

Design/Organization ... 1
  Organic and Inorganic ... 1
  Client Services ... 2
  Quality Assurance ... 3
  Data Management ... 3
  Sample Management ... 3
Management ... 4
  Measuring Capacity and Backlog ... 4
    Laboratory Capacity ... 4
    Backlog ... 5
    Managing Backlog—Slow Periods ... 7
    Reporting Capacity ... 8
  Tests the Laboratory Offers ... 9
  Tests, Backlog, and Profitability ... 9
  Equipment Costs ... 10
  Staff ... 11
    Cost of Staff ... 11
    Staffing ... 11
    Training ... 11
  Information Management ... 13
    What Information Gets Managed ... 14
    Manual Tracking Systems ... 15
    Semiautomated Tracking Systems ... 16
    Transfer of Data from Analytical Instruments to a Microcomputer ... 17
    Laboratory Information Management Systems (LIMS) ... 17
    Getting Data to the LIMS ... 17
  Reporting ... 18
    Reporting Formats ... 18
    Tiered Approach ... 19
    Your Product ... 20

Quality Assurance 20
Quality Assurance Project Plans (QAPjP) 20
Project Planning — Using the Checklist 21
Sample and Reagent Disposal 26
Treating Hazardous Waste 26
Listed Wastes — Some Advice 27
A Final Bit of Advice 28
References 29

Chapter 2
The Metals Laboratory
General Information — Matricies 31
Water 31
Soil 32
Sludges (Metal Hydroxides) 32
Organics 33
Air 33
Tips and Techniques — Sample Preparation 34
Sample Size and Detection Limit 34
Acid Selection and Acid Strength 35
Digestions for AA, ICP-AES, and ICP/MS 35
Digestions for Furnace AA 36
Heating Methods 36
Hot Plate 36
Microwave 38
Bombs 40
Sample Dissolution for Organic Matrices 42
Use of Blanks and Quality Control Sample Information 43
Special Cases — Mercury and Silver 44
Mercury 44
Silver 45
Metals Determinations 45
Standard Additions 45
Inductively Coupled Plasma 46
Nebulization 46
The ICP Torch and Flow Rates 48
Sequential and Simultaneous Instruments 50
Calibration 51
ICP Interferences 52
Sample Dilution 53
Internal Standards 53
Interferences and Background 54
Interference Check Samples 56
Flow Rate 57

Background Subtraction 58
Sequential Instruments 58
Atomic Absorption 59
Element Determination Using AA 59
Calibration 60
Maximizing Sensitivity 60
Interferences 61
Matrix Modification 63
Zeeman Background Correction 64

Chapter 3
The Organic Laboratory
Sample Preparation 67
Matrices — An Intoduction 67
Aqueous and Leachate Samples 67
Solids, Sludge, and Soil 68
Oils and Organics 68
Air 68
Sampling Heterogeneous Materials 68
Volatile Samples 69
Direct Injection 69
Purge and Trap 70
Headspace 73
Semivolatiles 74
Ultrasonic Extraction 74
Soxhelet Extraction 75
Liquid-Liquid Extraction 75
PCBs and Pesticides 75
Sample Determination 76
Gas Chromatography 76
Injectors 76
Solvent Tailing 79
Injector Liners 79
Leaking Septa 81
Leaking Ferrules 82
Foreign Material in the Injector 82
Peak Splitting 84
Columns 85
Acid/Base Behavior 88
Loss of High End 88
Injection Speed 89
Autosamplers and Sample Carry-Over 89
Gases 90
Carrier Gas and Chromatographic Resolution 91

Carrier Gas and Detectors 92
Carrier Gas Flow Control 92
Detectors 93
The Flame Ionization Detector 93
Electron Capture Detector 97
Photoionization Detectors 98
Electrolytic Conductivity Detectors 98
Compound and Mixture Identification and Quantification 98
Gas Chromatography — Mass Spectrometry 101
Low Response vs. the Internal Standard 102
Low Response 103
The Sample is not Getting to the Source 103
Leaks 104
Not Forming Sufficent Ions in the Ion Source 105
Analyzer 105
Detector 107
Scan Speed 108
Electron Energy 110
Ion Source Temperature 110
Instrument Zero 111
Tuning 111
Data Reduction 113
Compound Identification - False Positives and Negatives 113
Quantification and True Detection Limits 115
Avoiding Reanalysis 116

Appendix I
Example ASCII Files from GC/MS Data Systems for Export to LIMS 119

Appendix II
Preparing Quality Assurance Project Plans 125

CHAPTER 1

Laboratory Operations — Design and Management

The purpose of this chapter is to provide information about managing an analytical laboratory. Probably the biggest problem facing laboratory managers is how to manage their resources. By that I mean personnel, equipment, and sample backlog. Managing each of these is critical to the success of the laboratory manager and determines whether the laboratory will be profitable.

DESIGN/ORGANIZATION

Most laboratories are organized into a structure that mirrors the analyses the laboratory offers. Figure 1-1 provides a generic organization chart that fits many laboratories. Generally, the lab is divided into two main sections covering inorganic and organic determinations. Additional sections might be required depending on the specific analytical mix offered by the laboratory and sample load. For example, if your laboratory offers asbestos determinations you might have a separate microscopy section, assuming the sample volume for asbestos warrants it. How you arrange these elements (i.e., into a reporting structure) depends on your laboratory and the analytical mix offered. A brief description of the major elements follows.

Organic and Inorganic

The inorganic section normally handles inorganic sample preparation (i.e., digestion), metals determinations (usually atomic absorption, AA and inductively coupled plasma, ICP), anion determinations (ion chromatography), and wet chemistry (cyanide, sulfide, total dissolved solids, etc.). The organic section covers sample prepa-

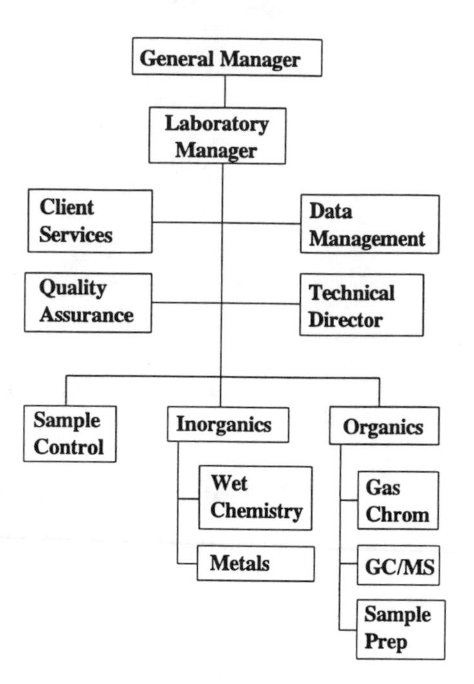

FIGURE 1-1. Organzation of a typical laboratory.

ration (extraction and dissolution), gas chromatography (GC; primarily pesticide and PCB analyses), and gas chromatography-mass spectrometry (GC/MS; volatile and semivolatile organics). These groups are the main production groups in the laboratory and represent the laboratory's front line in implementing the laboratory's quality assurance (QA) program.

Client Services

Many laboratories have adopted a structure that includes a client services group. This group is responsible for working directly with clients to book samples for analysis, setting schedules for data delivery, ensuring that the data provided meet client data requirements (e.g., making sure the client understands that you don't need full contract laboratory program deliverables for most projects — in other words helping clients define their data quality objectives), ensuring that the sampling team has the proper sampling containers and that they preserve samples correctly for shipment to the laboratory, negotiating prices, etc. This group also discusses any special requirements the client may have, for example, special data reports, diskette deliverables, nonstandard detection limits, applicable regulations (i.e., what regulations are driving the sampling and analysis project), etc. This group coordinates accounting with the financial staff and works with the client to obtain site historical information that may be helpful during the analysis phase. This group also provides day-to-day information to the client regarding problems or schedule changes.

Quality Assurance

The quality assurance group is responsible for setting up and managing the day-to-day quality assurance program for the laboratory and ensuring acceptable performance in external QA programs. This includes establishing the laboratory's quality assurance management plan and managing an in-house quality assessment program (i.e., performance evaluation samples, maintaining quality control charts for control samples, duplicates, blanks, etc.). The quality assurance group also monitors and maintains the laboratory's certifications and performance in external quality assurance programs. These would typically include quality assurance programs of states, the Environmental Protection Agency (EPA), and other standard-setting organizations.

The quality assurance group ideally reports directly to the overall laboratory manager regarding deficiencies and results of quality improvement efforts. The group independently reviews laboratory procedures to ensure that the data generated meet client needs and stays apprised of changes in regulations which affect the laboratory's procedures and quality assurance program (e.g., the mandatory use of Chapter One for Resource Conservation and Recovery Act [RCRA], etc.). This group must also work with production and client services to produce quality assurance project plans (QAPjP) when they are required for specific projects.

Often management is tempted to sacrifice quality for production. The quality assurance group must work with laboratory management to ensure this does not happen and that the laboratory has procedures in place to deal with production problems when they arise. The QA group may also be responsible for personnel training and making training recommendations to management.

Data Management

The data management group is responsible for maintaining records in conjunction with the sample management group and is usually responsible for putting final reports together and checking the data to ensure that method-specific QA and quality control requirements have been met. In this regard data management and quality assurance overlap. Unfortunately, when data are detected that do not meet QA specifications it is usually too late to implement cost-effective correction. This is discussed in the Information Management section later in this chapter.

Sample Management

The sample management group is responsible for receiving samples, ensuring that the samples are documented (i.e., have necessary chain of custody), notifying the lab manager and client service representative of problems (i.e., broken sample bottles, etc.), logging the samples in for analysis, and tracking the samples through the laboratory. This group may also be responsible for tracking laboratory capacity and sample backlog.

MANAGEMENT

As indicated above, the most important function of the laboratory manager is to manage resources — personnel, equipment, and backlog. The survival of the laboratory depends on how well these are managed.

Measuring Capacity and Backlog

In order to manage backlog, you need to have some idea of your capacity within the laboratory for each analysis the laboratory offers. You also need to know how much each analysis costs and what you can charge. In other words, you need to know the approximate amount of revenue and pretax profits you generate for each analysis. Now, how do you do that?

Most laboratories measure capacity differently; however, they all have the same goal. They are trying to determine how many samples can be analyzed and reported to the customer (i.e., billed) in a given period of time. This establishes capacity. I have had many discussions with laboratory people, engineers, etc., and everyone has a different definition of what constitutes a sample. The more important factor is how many times you can perform a specific analysis in a given time period.

To illustrate the problems encountered when discussing "samples", consider the following. When engineers or sampling people define a sample they normally mean a material that is taken from a specific place at a given time. A groundwater sample may be taken from a specific well (i.e., at a particular depth), on a given date. Now that "sample" will be divided into several jars and containers before it is transported to the laboratory. The number of these "sample" containers is dictated by the number of analyses that will be performed. As far as a project manager is concerned the groundwater from the well represents one "sample". In the laboratory it may represent anywhere from one to ten (or possibly more) samples. When you think about it, the term sample has no real meaning for the laboratory as far as capacity is concerned. Again, what you need to know is how many times you can perform a specific analysis in a given period of time. That is the capacity for a specific analysis.

Laboratory Capacity

The true number of analyses you can perform will be a percentage (usually less than 50%) of theoretical. Let's illustrate this using ICP atomic emission spectroscopy (AES) as an example. Suppose you have a sequential instrument and you are analyzing a standard suite of metals. Let's say this takes 5 min for each sample. Theoretically you can analyze 20 samples an hour. For a 5-day work week and one shift per day you can analyze 160 samples per day or 800 samples per week. The 800 total is an upper limit, assuming everything works correctly. Now, we know you have to warm up the instrument, verify the calibration (or perhaps recalibrate), verify correction factors, etc., each day. Let's say this takes about 1 h each day (i.e., the equivalent of 20 samples). So our sample number per week is down to 700 because we lose an hour every day. Now, every 10th sample is followed by a check sample and a blank just to make sure the instrument calibration is alright. That means 2 samples out of 12 are not paying samples, or 17% of the capacity. Subtracting 17% from 700 we get 580. You also have to analyze a laboratory control sample with every batch or at least every 20 samples. So we lose another 5% for that, bringing our total down to 550 samples. Now, on every non-QA sample we either dilute the sample and rerun it to ensure there are no interferences or we perform a bench spike. This cuts our sample number in half. We are now down to 215 analyses per week. So, you can see our efficiency for ICP-AES is only 215/800 or about 27%. Keep in mind that your efficiency will be this high assuming there are no other problems — that is, no instrument maintenance or sample-related problems. Suppose one out of every ten samples is multiphase and that you lose about 5% of time for instrument maintenance. This cuts your efficiency still further down to about 170 billable samples per week. So our overall efficiency will be about 170/800 or 21%, only 34 samples per day.

The laboratory manager should go through this exercise with every piece of equipment or sample processing operation in the laboratory to get some idea of actual analytical capacity. Establishing analysis capacity is the first step in managing sample backlog. Using the example above, the laboratory manager can plan on 34 ICP-AES samples per day. Now, it is the manager's job to keep 34 samples a day (on average) coming in the door. This is not an easy job, given that sampling projects are delayed or canceled, the number of samples that show up may be higher or lower than agreed, instruments go down unexpectedly, etc. Therefore, there are always peaks and valleys in demand. This is where things get tricky. If you try to operate with too large a backlog you run the risk of missing holding times or you may not be able to meet promised turn-around times. If you miss turn-around times customers will feel that your lab is unreliable. If you miss holding times you may be responsible for the cost of resampling. Either situation is undesirable. So, how can you establish reasonable planning backlogs for your laboratory operations?

Backlog

A planned backlog for each laboratory operation can be related to a planning horizon and sample holding times. First you need to know what your capacity is for each analytical unit operation. Let's use the same example as above and assume we have a capacity for ICP-AES of 34 samples per day. The next thing you need to know is how many samples you have in-house for analysis. Let's say you have 400 samples in-house requiring metals analysis. Next you have to pick a planning time, for example a week, a month, a specific number of days, etc. Let's say you arbitrarily select 30 days for this period (i.e., an average month). Keep in mind that there are only 21 working days, on average, in this 30-day period. Your capacity of 34 samples a day means that you can analyze 714 samples during that interval. After your 400 samples in-house you have additional capacity for 314 samples. You need to book an additional 314 samples over the next 30 days in order to operate at full capacity. Actually, you will only need about 16 days to complete the 400-sample backlog. So, that means you really need to book the work within about the next 2 weeks. To help with booking samples you might offer discounts to customers who can deliver samples within this time period. Alternately, if instruments require periodic maintenance, you can schedule down-time during periods when you are below capacity. Remember, your profitability depends on how well you utilize your capacity.

Most laboratories receive samples daily so you should go through this exercise at least once each week. During peak demand time (e.g., summer and fall when the weather is more conducive to sample taking) you may need to do this every day.

Backlog management is a critical function that should be performed by the laboratory manager or designee. As pointed out above, if the laboratory is operating at less than 100% capacity, you lose the opportunity for additional revenue. So what can the laboratory manager do to ensure that there will be a backlog? Unfortunately, the answer is not very much, but there are some strategies that good managers use to help with the backlog problem.

As a laboratory manager, one of the best things you can do to ensure backlog is to establish a diverse group of customers and ensure that no one customer accounts for more than a small percentage (e.g., 10%) of the laboratory's capacity. Consider the following two situations. A lab that has three major customers, each accounting for about 30% of the lab's business, and a second lab that has several hundred customers but no one customer accounting for more than 10% of the lab's business.

The lab with three major customers also has about 20 small customers that account for the remaining 10% of sales. As long as the three major customers provide samples at a steady rate the lab can operate profitably and meet requirements for turn-around and quality. If any of the three major customers has a slow down or a change in sampling, the laboratory will have a problem in utilizing capacity. On the other hand, if one customer has an immediate need for additional work, the lab can meet needs for additional capacity by sub-contracting some of the work out, adding personnel, adding shifts, etc. The real problem occurs when customer needs go down. The laboratory will be underutilized and ultimately lose money. In this case, the lab may need to lay off personnel or scramble for additional work to cover the shortfall.

The laboratory with hundreds of customers is in a much better position to weather a slow down by one or more of its customers. If one customer does not send samples when expected or has a slow down, the shortfall will more than likely be met by other customers. The laboratory has effectively lowered its risk by cultivating many different customers. Furthermore, the laboratory has actively sought customers in the public and private sectors to further lower overall risk. This laboratory manager realizes that even if he has many customers, but a large number of them depend on government funds, the risk to his lab would be higher, since budget shortfalls are common in government agencies and could be devastating to his laboratory.

The basic rule to follow is to have a plan to attract as many customers as possible and ensure they are from a diversity of industry, government, and academia.

Another approach, although riskier, to ensuring backlog is to over-book the lab. An analogy is airline flight booking. Airlines will normally over-book flights since there is a high probability that one or more passengers will miss the flight or cancel their reservation at the last minute. Through the use of statistical techniques, the airlines have established rules by which they can over-book and still make a profit. Problems occur, however, when more than the expected number show up for the flight.

Laboratories are in exactly the same situation. The lab has a specified capacity for each analysis it offers. The manager should over-book that capacity to ensure optimum efficiency; however, there will be times when the numbers of samples will out-strip available capacity. In this case the lab can handle the situation the same way the airlines do. Inform customers of the delay, ask for their cooperation in rescheduling the work this may include some incentives for customers whose work is delayed, subcontracting out work, etc.

You can see that there are two strategies you can use — diverse customer base and over-booking — to ensure a backlog. There are a few other approaches dealing with contractual mechanisms that are now being discussed in the laboratory community; however, these have not yet gained acceptance. The laboratory can negotiate with customers who book significant capacity to establish a minimum payment if samples are delayed by more than an agreed-upon time or offer customers incentives to get samples to the lab within a specified time frame. This is how it works. If a major client anticipates a large project to start on a given date with samples arriving at the lab over a 2-month period, the lab can price the work to include a minimum payment (like a down payment) to hold that capacity. If samples do not arrive within the anticipated time frame, the lab may adjust charges, within prenegotiated limits. Thus, the customer will get the best price by working with the lab to schedule arrival of samples. If the sampling schedule slips, the customer is asked to help share the financial risk.

Managing Backlog — Slow Periods

All laboratories experience peaks and valleys. The best way to handle peak demand and low demand times is to first stay as flexible as possible within the laboratory, and second to try to manage your customers. The best approach is to maintain a high degree of staff flexibility. For example, if the metals laboratory is under-utilized you may be able to use those personnel to augment data review and reporting staff. You might be able to use GC and GC/MS operators interchangeably. The ideal situation would be to allow infinite flexibility to move personnel to any position in the laboratory and have them function effectively, and as efficiently as possible. Of course this is not entirely possible. However, the point is, the more flexibility you have with staff, the better. This will give you a better chance to keep everyone working in areas that ultimately result in billable analyses.

Managing customers, although not always possible, is probably the next best method for balancing the work load through your laboratory. Customer management means working with customers to establish times when samples will be taken and shipped to the laboratory. For example, when you work with major clients, you might be able to delay sampling to ensure laboratory capacity. You can do this by offering price or turn-around time incentives. This approach works when the demand for services is high and supply is low. For example, the demand for radiochemical analysis might be high and supply limited. When this is the case you can influence sampling dates when samples will be drawn for radiochemical analyses. Although customer management is desirable from a laboratory point of view, your customers might think otherwise. If another laboratory has available capacity you could lose business if you try to manage the customer.

Reporting Capacity

As discussed above, most labs are organized into specific units. There are metals analysis units, pesticides, GC/MS, etc. These units represent the heart of the laboratory. Report generation, however, is probably the most important unit in the laboratory. Without reports the laboratory doesn't get paid. Unfortunately, reporting is an area that is easy to overlook when estimating overall laboratory capacity. You should go through the same exercise to determine the laboratory's reporting capacity as you did for all of the other operations of the laboratory. Depending on the types of reports you must generate, you will likely find that reporting is a rate-limiting step in your laboratory. This was evident in the early days of EPA's Contract Laboratory Program (CLP). EPA's CLP has literally hundreds of reporting requirements. These requirements involve providing certain types of information and then packaging that information in highly specialized formats. When they started working in the CLP program most laboratories found it difficult to generate reports in the required format in the required time. As time went on, most of the labs in the program became extremely competent at creating CLP "Data Packages" in the time required. The point is that it takes time to create reporting systems and debug them and this is one area that is easy to overlook in your laboratory. I think it's safe to say that most laboratories can generate reliable data within the required holding times; however, many labs find that reporting in the time required is the most challenging part of the process.

As indicated above, you should evaluate your ability to generate reports just as you would evaluate any production step in the laboratory. Every type of report that you generate is like a different test or method. For example, to generate a complete CLP-type data package for 10 samples might require 16 person hours. Depending on the amount of data review you use in your laboratory this time may be more or less. At any rate, you have to know how many different type reports you can expect to generate with available resources. Since report generation happens after analysis this is one area where you can use additional personnel when you are not at full analytical capacity.

Tests the Laboratory Offers

Every analysis you offer has an associated cost for the laboratory. Every time you add a new test or procedure you increase your costs of doing business. A new test or method might require a new piece of equipment, additional staff training, new reporting systems, additional sample tracking systems, etc. Often the amount you can charge for a specific test is dictated by the market. For example, if a standard organic volatile analysis (water) costs, on average, about $250, it will be difficult for you to charge $350 or more for the same analysis, regardless of how much that analysis costs you. Furthermore, customers are often unwilling to split samples between two laboratories to take advantage of prices for specific analyses. This means that you will have to offer a core set of analyses which will include some that have a very low profit margin or may even involve a loss.

For example, some labor-intensive tests are expensive for the laboratory, however, they may be relatively inexpensive on the open analytical market. Tests such as total dissolved solids, total suspended solids, total solids, cyanide, sulfate, etc., are likely to have very low margins. Thus, to maximize profits a laboratory should drop these tests from their offerings in favor of more profitable tests (e.g., GC/MS). Unfortunately, customers may routinely require these low margin tests along with the higher profit tests. Since customers are generally unwilling to split samples between laboratories you are forced to offer low margin tests, even though they are unprofitable — or even generate losses, to remain competitive overall.

Tests, Backlog, and Profitability

When making decisions about whether to offer a new test you should establish a realistic cost per sample given the anticipated demand (backlog) and what you might reasonably charge for the analysis. Be sure to consider staff time, equipment usage, additional overhead, additional reporting requirements, additional inventory costs associated with reagents, etc. If your calculation indicates that you will turn a reasonable profit (20% or better) offer the test. If the calculation shows less than 20% ask yourself how much business you might lose if you didn't offer the test. If you stand to lose more than a few percent of your total business you might want to offer the test to protect overall competitiveness. In this situation, probably the best solution is to farm this test out to another laboratory. Thus, you send low profit work to another lab and keep the higher profit work for yourself.

Every time you add a new analytical method (test) you increase your costs. Expenses include staff, training, supplies, tracking, reporting — everything. When you add a new method/test it has a ripple effect in the lab. First, you have to make sure that sample receiving understands what the new test is, how to store the samples, what sample containers are appropriate, preservation techniques needed, etc. If you will be storing the samples for any length of time you will need to have space allocated. If the samples must be refrigerated you should check to make sure there is adequate space to cover the anticipated volume. If there are special reagents, these need to be on hand. This includes standards. Remember — storage requirements also apply to standards. You have to be sure to have the right instrumentation to offer the test. This seems obvious; however, you should make sure whether you need additional equipment or if you can convert existing equipment over for the new test. If you opt to change over existing equipment you have to consider lost time for both tests (i.e., the one being changed from and the one being changed to). Depending on how often you change back and forth, this can affect your overall costs. If you buy a new piece of equipment, amortization should be included in the analysis price. When you add a new test you must consider whether there are associated training requirements. Some contracts require a certain experience level with a given method.

All tests have associated reports. Can this new test be included on a standard report that you already generate or does it require a new type of report? The answer to this question will affect your costs. How much QA is appropriate for the new test and what does that cost? You need to think about waste problems associated with offering new tests. Laboratories that are considering getting into mixed waste testing or dioxin testing should consider what they will do with their wastes before offering these tests. Since there are limited disposal options for dioxins and mixed wastes your decision to offer these tests may be influenced by waste disposal options available to you.

The bottom line for all of this is that you need to know *exactly* how much a test costs before you decide to offer the test and whether there are potential "show stoppers" in associated factors. You also have to consider whether you can maintain a backlog for the test. If you can't you need to multiply the cost by that backlog factor. For example, suppose you calculate that all of the costs associated with offering ICP-AES (including all of the QA/QC samples) comes to about $200. If you can only keep the metals laboratory busy about 80% of the time then you should multiply $200 by 1.2 to get $240. This should be the price charged by the lab for this analysis. Now it's true that for 20% of the time you will not be generating wastes, reports, need chemicals, etc., so the 20% add on may be a little high; however, it's better to be conservative until you have enough experience with a particular test to know exactly where your costs are and establish pricing that reflects those costs. For additional information on this topic see Taylor and Amano's article in the *Journal of Chromatographic Science*.[1]

Equipment Costs

I find that it's always a good practice to buy used equipment preferentially over new if the equipment is in reasonably good condition. This approach lowers the fixed costs for the laboratory and is the least expensive way to enhance your capabilities. Typically, you can save between 50 and 80% of the price of a new instrument. For additional information see Michael Martin's article in *Environmental Lab*.[2]

Staff

Cost of Staff

One rule of thumb I've developed over the years is that most companies, laboratories included, must generate about $100,000 in sales, on average, for every employee. A ten-person laboratory should have about $1,000,000 in sales to be profitable. This does not mean that laboratories with sales less than $100,000 per employee will be unprofitable. It just means that you will probably realize the best return on investment when you maximize this value.

Staffing

The staff of most laboratories encompasses a mix of people. This includes technicians, personnel with bachelors degrees, and those with higher degrees. Most environmental laboratories will be made up primarily of technicians and personnel with bachelor degrees, with perhaps a few higher-level personnel for more sophisticated equipment if the laboratory has any (e.g., GC/MS instruments, etc.). Generally, you find degreed personnel operating the atomic absorption spectrometers, the inductively coupled plasma atomic emission spectrometer, and gas chromatography-mass spectrometry equipment. Technicians normally work in sample receiving and preparation.

While it is important to have qualified and well-trained people in the measurements laboratories, many of us are coming to realize that sample receiving and sample preparation are two of the most important areas in the laboratory. Even though these areas are very important to the overall function of the laboratory, they are usually staffed with junior personnel.

Sample receiving should have a mix of junior and senior personnel. The sample matrix will often dictate appropriate analysis methods. Personnel that are skilled in evaluating matrices (e.g., by visual inspection) and determining appropriate analysis techniques are invaluable. They can alert the respective sample preparation and analytical personnel that they may need to change equipment configuration or set up special analysis conditions in advance and recommend the correct size of sub-samples for subsequent analysis. This often requires special skills in determining whether homogenization or compositing is needed to get sub-samples that are representative of the sample. Therefore, a mix of junior and senior personnel in these areas is desirable.

Training

Most laboratories have in-house training programs where they assign junior-level personnel to work with a senior technician or chemist. This approach is the best, low cost alternative available; however, it does little to really up-grade the laboratory's capabilities. The person doing the training trains people to do the task the way it's always done. While this is effective in getting new people indoctrinated in how the laboratory works, it does little to enhance the overall productivity of the laboratory.

Now, if a 1 week training course increases a person's productivity by only 5%, you will realize a full return on your investment in about 1 year. (Since 5% of $100,000 is $5000 and assuming 1 week of training will cost about that amount. If the person's productivity goes up by 5% you will recoup the $5000 in one year.) The 5% is generally considered conservative.

There are several laboratory trade organizations in the United States. Most of the major laboratories belong to at least one of these organizations. These organizations are the International Association of Environmental Testing Laboratories (IAETL), headquartered in Arlington, Virginia; the American Council of Independent Laboratories (ACIL), headquartered in Washington, D.C.; and the Association of Laboratory Managers (ALM). Just about every meeting sponsored by these associations devotes several papers or meeting subelements to training. These papers, as well as contacts within these organizations can provide a good source of information about training programs.

The money that can be used for training is in competition with up-grading and replacing equipment, adding new services (analyses), travel and attendance at conferences, etc. Everyone I've ever talked to thinks that training is essential to the success of the laboratory. However, most labs invest little in training programs, unless they are required to do so. The reason for this is as follows. The laboratory testing market in the United States is growing, even though growth in the 1990s is not as fast as in the 1980s. There are about 300 environmental laboratories in the U.S. and about 1000 commercial testing laboratories overall. Laboratory personnel generally know about other commercial laboratories (i.e., where they are, who works there, etc.) and often have contacts at these facilities. Contacts are made at professional meetings and through open literature sources. All of these facts make it relatively easy for one laboratory to hire staff from other labs, often by offering higher salaries, better benefits, better working conditions, more responsibility, etc. The turnover in technicians and professional staff at commercial laboratories can be high (as high as 20% per year — which affects the overall performance of the laboratory in the market). The laboratory manager is faced with a dilemma. Why should you pay to train personnel who may take a job elsewhere before you realize an adequate return on your training investment? A very difficult situation.

Instrument manufacturers can be a good source for training. Often the manufacturer provides training courses for their instruments at a reasonable cost, typically about $1000 for a 1-week training class. If you are purchasing an instrument you can often negotiate for a training class as part of the price. Of course you will also have the travel expenses of sending a person for the training. Instrument sales representatives often have a budget that can be used, at least in part, to provide training, both off- and on-site. They can also arrange for people from their marketing department to provide what amounts to training. The only problem with arranging on-site training is that your staff can be interrupted if something goes wrong in the laboratory and miss some of the instruction. When you talk to sales representatives ask about training they might provide. If you arrange for on-site training try to do it such that your staff cannot be interrupted. For example, you can book a room at a local hotel for part of the training class and then have a hands-on session at your laboratory later in the day.

Some laboratories are experimenting with training agreements. If the laboratory pays for a training course, the employee agrees to stay on with the laboratory for an agreed upon time. For example, every week of training can equal 6 months. Thus, after a one week training course that person is obliged to stay on for at least 6 months. The government uses similar agreements to retain personnel after training.

The laboratory should have a system in place for keeping employee records up to date regarding qualifications and training. This is important since many contracts require that analysts performing specific tasks have a specific level of training, education, and/or experience. For example, a contract may require that GC/MS operators have a bachelors degree plus 6 months experience in operating a GC/MS system. Personally, I think that most of these contract requirements are unnecessary; however, if you want to bid on these types of contracts you have to be able to show you can meet the contract requirements. An easy way to do this is to include education, experience, training courses, etc., in every employee's personnel folder. In the front of the folder have a checklist. The employee should check off the appropriate boxes to indicate training (including required health and safety training) and education. This is also a good way to keep up with Occupational Safety and Health Administration (OSHA) requirements. For example, part of the checklist can be devoted to whether the employee has read certain MSD sheets, has had safety training, etc.

Information Management

Sample tracking and management systems are an integral part of any laboratory. These systems are not only important to ensure that the laboratory is working efficiently but also to provide information for cost accounting. As a laboratory manager you need to keep track of when samples come in, what analyses are required, required data quality, and reporting requirements. You also need to make sure that there is a linkage between the sample tracking function and accounting. In other words you need to make sure that when a sample analysis is completed and the report is shipped to the client, an accurate accounting of the services performed is also sent to the client. Some laboratories also use sample tracking information for ordering supplies.

There are lots of ways to handle sample tracking, accounting, and purchasing information. The best way is to have one system that does all three. Unfortunately, most labs do not start with such an integrated approach. Once tracking and accounting systems are in place and operational, they are difficult to change. In this section I discuss various methods of organizing these functions using manual and automated systems.

Regardless of whether you use a manual or automated system, you should probably set up all information that the laboratory generates around the idea of an analytical method.

What Information Gets Managed

Managing information based on a sample and its associated analysis is the basis for any laboratory information management system. This information is divided into three general categories. There are sample specific, analysis and compound, and quality assurance types of information that need to be managed.

Sample specific information, which can be referred to as the sample (specific) header, includes the following:

- Client sample identification number
- The laboratory sample identification number
- Contract number, client name, sample delivery group, case number, etc.
- Analysis method (one record for each method), fraction, etc.
- Date sample was taken, received, shipped, analyzed, extracted or prepared, and data verified
- General sample information, such as matrix, pH, percent moisture, sample weight used, preparation batch number, etc.
- Type of sample (i.e., blank, duplicate, calibration, spike, QC, etc.)
- Related sample numbers, such as the associated blank, spike, duplicate, calibration, standard, etc.

The second type of information deals with the analysis method and the compounds covered by that analysis. Here is where you describe the basic information about how the analysis works:

- The name of the analysis or method number
- The compounds or elements which are covered by that test
- The result (e.g., concentration) for each of the compounds found

The quality assurance information is associated with each method and for each associated compound or element. This is where you build in the basic quality assurance for each method or test the laboratory performs. This information is what ties the information management system together as follows:

- For each compound you list the associated reports (i.e., tests)
- For each compound and report combination (test) you include relevant test limits you need to perform

Each of the three types of information is related. When the sample is logged in a record is created for each analysis. The header information is filled in for each record. As the analyses are completed the analytical data are collected and tested. You track samples by looking only at the header information. The header contains dates and times the sample was prepared and analyzed. After analysis you can check the analysis results and make sure that all required information is available.

You do this by taking each sample record and generating the reports you want. Data are recorded for each compound detected. There is no need to record data for undetected compounds, since the absence of a result means the compound was not detected. Using GC/MS as an example, you may want to generate a holding time and initial calibration report. For holding times all you need is the header information and to check whether all of the samples were analyzed and/or extracted within the time limits. To generate the response factor report you use the header file to identify the initial and continuing calibration files associated with each sample. All compounds show up on a response factor report; however, only a few compounds have associated limits. These limits will be listed with each compound.

In summary, the method you use defines the compounds and the types of reports you generate, the reports define the compounds that get tested and the limits of the tests. In a very real sense, the reports you generate define the data base you use. The approach outlined above makes extensive use of indices to manage information. Now, consider different ways you might want to implement this type of information management system.

Manual Tracking Systems

Let's discuss sample tracking and data management in the laboratory. First, we'll consider manual tracking systems. When a laboratory first gets started, it may have only a few people and there is little need for a sophisticated sample tracking system. Most of the people in the lab know what has to be done and understand the order of sample analysis. In this case, a simple paper tracking system is the most reliable and least expensive. When the samples arrive the sample numbers are transcribed onto a work sheet for each analytical operation. Of course, all of the header information discussed above needs to be recorded, but the work sheets only need the sample numbers, dates, matrix, and methods. For example, if a sample will be analyzed for metals and volatile organics, the person receiving the samples will list those samples on work sheets for the persons who will analyze the samples for metals and volatiles. The work sheets are then given to the technicians and analysts and the laboratory manager should get a duplicate. The analysts schedule the work for their individual laboratories by using the work sheets. They should also keep the laboratory manager informed of work progress. This should occur on a routine basis.

Each laboratory section generates hardcopy data. An easy way to know when the analytical work is completed is to use a set of mail slots in an office or room designated for report assembly. Many labs use the room where the copier is located for this, since it often involves copying. When a group of samples is logged in, a set of mail slots is designated with the sample numbers and slots are assigned for each analysis (method) that is performed. At the end of each day, the analysts take their hardcopy results and place them into the slots designated for each sample analyzed. For example, if metals and volatiles are designated for a given sample there will be three or more mail slots with that sample number and an indication of the data that goes into each slot. In this case one slot is allocated for the ICP data, one slot for the volatile organics data, and one slot for the mercury analysis. All you need to do is to check the mail slots from time to time to see how work is progressing. Slots that are filled indicate that the analysis is completed. Empty slots mean that the work is still in progress. This represents a very simple, but highly effective tool for the laboratory manager to keep track of the number of samples in the lab and how fast the work is progressing. It also can quickly show when bottlenecks are developing.

When the slots are filled for a given sample or sample group, a report can be assembled and the data shipped to the customer. Many labs use colored tape to section off the mail slots into sample delivery groups. When all the slots are filled inside of a colored boarder, the report is generated. With this system, the analyst is responsible for ensuring that all of the required information is placed into the slot. For example, if the customer requires QC data, such as the initial calibration, continuing calibration, spikes, duplicates, blanks, etc., the analyst must include this information with the sample data or reference a previous set of reports where the data can be found. This may be accomplished by including a QC slot with each analysis type or by simply including it with one or more of the samples. In either case, you have to rely on the analyst to ensure all of the proper analytical information is included.

Many laboratories, even ones that are fairly automated, use the mail slot approach to keep track of work going through the laboratory and when all of the data have been generated so that reports can be assembled.

Semiautomated Tracking Systems

Most of the laboratories I've visited are what I would call semiautomated. That means some functions are tracked electronically, while others are manual. Keep in mind that one objective of a good laboratory manager is to try to automate as much as possible. In this case, sample tracking information can be entered into a data base or spread sheet which the laboratory manager and the individual laboratory section managers use for scheduling work and tracking samples. This type of system amounts to the same thing as the manual system, except you use computers to generate the work sheets and you can up-date and modify the work sheets more easily. Generally, when a laboratory first starts to use computers to track sample information (i.e., the header) they don't think about linking it to the method, compound, and report information. Making this linkage is important and you can avoid future problems if you keep in mind how the different types of information are interrelated and then use software that indexes everything to keep it all straight.

With this type of system it becomes easy to begin to look at some fundamental laboratory statistics. This is valuable since it allows the laboratory to measure how well it is performing in certain critical areas such as:

- Turn around times
- Instrument utilization rate
- Revenue per instrument
- Revenue per employee and per employee type, etc.

Obviously, as a laboratory begins to grow this type of information is valuable for planning acquisition of both equipment and personnel. You should also be able to track the amount of re-work being done in your lab.

Transfer of Data from Analytical Instruments to a Microcomputer

Most instruments have the capability to export summary data (i.e., results lists) in ASCII format. These data can be easily transferred to a PC and imported into any one of several popular programs. You can import a quantification report into a text editor program like Word Perfect® or into a spreadsheet like Lotus 123® or into a data base program like Dbase®. It all depends on what you want to do with the data. If you want to role the data up into reports with text you might chose Word Perfect®. If you want to perform calculations you might use Lotus. If this is the case you will most likely need to write a MACRO to read the files and parse the data into a spreadsheet that will perform the calculations or display the data you want. If you want to combine the summary results data with general sample information you might want to use Dbase. You can write MACROs in Dbase to import data from the ASCII files and parse these data into a data base. Each of these approaches will work and many labs have employed all of these and combinations of them.

Laboratory Information Management Systems (LIMS)

When the sample volume in the laboratory is more than about 100 samples per week you will need to consider higher levels of automation. There are several laboratory information management systems (LIMS) commercially available today. I estimate that about half of the laboratories with commercial systems use a Perkin Elmer (or PE Nelson) LIMS. After PE, you will find that some labs use Hewlett Packard, Radian SAM, or other commercially available systems. Many labs, once they have gotten big enough, decide to build their own LIMS. Building a LIMS is a very expensive proposition. Usually, you will have to hire one or two programmers for about 2 years to develop most of the code you need. You will also have to invest in hardware. Most laboratories that take this route usually spend about $2 million (or more) for their LIMS system. Given the fact that you can buy a fully functional LIMS for about $50,000, I don't quite understand why a laboratory would want to tie up that much money in a home-made LIMS. I've asked several laboratories this exact question and have never gotten a satisfactory answer. Generally they say something like "we have a unique system and none of the commercial LIMS fits our needs". If I were in this situation I would change my "unique system" to conform to one that an off-the-shelf LIMS could handle and pocket the difference.

The section below is intended to provide a brief introduction to LIMS; their pros and cons and where you might go for additional information.

Getting Data to the LIMS

A LIMS is not usually used to store raw data. Only summary reports are sent to a LIMS. Let's illustrate this by considering GC/MS data. The GC/MS data system collects the raw GC/MS data — typically mass, intensity, and scan information (see Chapter 3 of this book for a more detailed description of GC/MS). The output from the GC/MS data system can be summarized in an ASCII file with the results from quantification of target compounds. All modern instruments are able to produce a file of tabular results. This file can then be transported to the LIMS for additional processing and reporting. An example of these reports is provided in Appendix I.

These reports are transferred to the LIMS via a file transfer program. If the LIMS and the instrument data system can use PCs as a terminal, you can accomplish the transfer by copying the data from the instrument data system to the PC and then from the PC to the LIMS. Many instruments today are using (or have the capability to use) a PC as the instrument data system. If this is the case the ASCII data are already on a PC and it can be up-loaded to the LIMS. A LIMS can also be tied directly to an instrument through a local instrument controller. It all depends on the instrument configuration and the LIMS you are working with. The problem isn't so much getting data to the LIMS. The problem is what to do with the data once you get it there.

Reporting

Reporting Formats

When it comes to reporting data, just about everyone has their own format. Actually, it's not quite that bad, but almost. The EPA has established several reporting formats through its contract laboratory program (CLP). Generally, states will specify reporting formats if they are contracting for analytical services. This is the case in New Jersey and California.

New Jersey has recently developed a reporting format for its contract laboratory program. They have a regulatory format, which is used for RCRA work and a different format for Superfund work. For additional information on these formats you can contact Andy Fishman at the New Jersey Department of Environmental Protection and Energy, Office of Quality Assurance, CN-424, 9 Ewing St, Trenton, NJ 08625–0424.

Generally, most states do not require reporting formats for data they don't pay for. This being the case, you should get some idea of what type of report and associated QA information your clients expect. This means that the laboratory, in conjunction with the customer, must decide on how much data and which data to report. At a minimum you will report the results for each sample analyzed for the specific analyses requested. The associated Quality Assurance and Control (QA/QC) data are where you have to decide what to report and how to report it. Since most commercial clients do not require all of the QA information, the laboratory can eliminate some or all of the QA/QC data. For example, you normally may not report data for calibrations, spikes, etc., unless specifically requested by the client. This point is the source of consternation at several labs that report some or all of the QA/QC data, since they feel they may be at a disadvantage because competitors do not provide the same level of detail in their reports. From what I've observed, I think the best advice is to report the absolute minimum that meets the customer's requirement. If you give them more information than they ask for it just gives them (and potentially their regulators) more things that they can question you about later.

The CLP reporting format is probably one of the best known. It specifies how to report data for each sample as well as how to report a myriad of quality control and quality assurance information. The CLP has changed reporting forms over the years and each time a change was made the laboratories had to make adjustments to their reporting software or rely on third-party software vendors to stay current. That's the bad news. The good news is that most of the changes dealt with reporting *format,* not substance. That means there are differences to be expected in *how* you report, not *what* you report. For example, you might be required to report certain information on one form but not on another. At a later date you might be required to duplicate certain information from one form on another. You're not being asked to report anything new — just the way the information is presented. If you have a flexible reporting system that captures the analytical data, as well as all of the QC data, you will be able to stay ahead of most reporting changes with a minimum of disruption. The point here is to stay flexible.

As we will discuss shortly, the EPA and several states have adopted standard reporting requirements that are based on EPA's contract laboratory program. These formats are designed around the standard procedures used by EPA and specify, in great detail, all of the required reports and even the order for assembling the final document. Most of us recognize that this has both positive and negative aspects. A standard reporting format is desirable since it lays out all of the requirements the laboratory must adhere to. This makes the laboratory's job easier in that you don't have a myriad of different formats that constantly change. On the other hand, if the number of specific reporting requirements is large (like the requirements used by EPA's contract laboratory program) then you will generally be over-reporting, that is, reporting information that the client neither requested nor needs. In fact, you may end up creating problems for yourself by reporting data to commercial clients in CLP format.

Since there are no hard rules for reporting much of the data you generate, the approach used by most labs is to design a standard data package (deliverable) that will meet the needs of most clients. If the client needs more information the laboratory might indicate that there will be additional charges or a surcharge to cover the additional reports.

A Tiered Approach

Many laboratories have adopted a tiered approach for reporting data to commercial clients. These tiers are generally based on guidelines established by the EPA. Basically, the way it works is as follows. There are four or five tiers (levels). The first tier is field test data. This might include pH measurements taken at a well, total chlorine measurements, etc. The second tier would be data generated using standard methods; however, without the normal level of QA/QC associated with that method. For example, we have generated tier two ICP data by using a single calibration point. A single calibration standard is analyzed, say at 5 ppm and then the samples are analyzed. If the metal is found in a sample, the sample is diluted to bring the concentration near 5 ppm. The result is then reported based on the dilution factor needed to produce the 5 ppm reading. In this case we generate data without following the QC specified in the standard ICP method. These results are called Tier II data.

There is quite a bit of confusion about what Tier III data are. Some feel that Tier II data are those generated by a laboratory procedure which has an associated standard operating procedure (SOP). The definition I've encountered most often, however, is Tier III data are the same as Tier IV data, without reporting all of the QC data. Tier IV data is defined as a full CLP-type report.

As you know, the CLP reporting format specifies a number of quality control parameters that must be met and reported. For example, it specifies how many calibration points are needed, their concentration, criteria for initial and continuing calibrations, etc. All of this information is reported on one or more forms in the standard CLP package. Thus, Tier IV is defined as the CLP. Tier III is the same as Tier IV in that all QC criteria and procedures are the same as Tier IV; the only difference is that you don't provide the full data package when reporting Tier III. For example, you would not normally include all of the initial calibration data, data for tentatively identified compounds, etc. All of the information maintained at the laboratory is the same for Tier III as for Tier IV; it's just not all reported.

For your laboratory you should establish what you mean by Tier III and Tier IV. Not all laboratories define these the same way. In summary, the definition we have found most prevalent is that Tier IV is equivalent to CLP and Tier III is the same as Tier IV except not all of the QC data are reported in the standard reporting package.

Your Product

The product from an analysis is a piece of paper with the results of that analysis. That piece of paper is a reflection of your laboratory and many labs go to great lengths to ensure that their product is not only technically correct and meets the customers' needs, but that it also looks and feels good. When reporting results for commercial clients, laboratories will often use a better quality paper and different reporting formats than are used (required) by federal and state governments. Most laboratories provide reports attached to a cover letter (bond paper with laboratory logo) which briefly describes the results and any problems encountered (e.g., some data may be delayed, some of the spiked or duplicates were outside of accepted ranges, some sample containers were broken on arrival, etc.) and the steps taken to correct the problems. The report is printed on a high-quality paper.

Quality Assurance

Quality Assurance Project Plans (QAPjP)

The EPA has proposed that quality assurance project plans (QAPjPs) be required for all projects where environmental data are collected. Several laboratories have developed checklists and generic QAPjPs that can be filled in to serve as a QAPjP if a client does not have a QAPjP for the project. We have already covered a good deal of the information needed in a QAPjP earlier in this chapter. Also, Appendix II describes a generic QAPjP that you can use.

Probably, the most important information you need to develop as part of the planning process is to understand how the data you generate will be used in decision making. That is, you need to define the decision that will be made with the data you are developing. The decision will, in turn, drive required method precision, accuracy, detection limits, etc.; and ultimately the analysis method. A decision might be to determine whether a waste is hazardous because of its lead concentration. If the concentration of lead in an aqueous sample is equal to or greater than 5 mg/l, the material is a RCRA hazardous waste. In this example, you don't need high precision or accuracy if the concentration is below about 0.5 mg/l or above 50 mg/l. You would like to have a method that could accurately and precisely determine whether a sample had 4.9 or 5.0 mg/l.

The process of establishing the decision and the decision rule is sometimes referred to as the data quality objective (DQO) process. The overall DQO process is fairly involved and beyond the scope of this chapter. However, the point is that you will, at some time, have to work with your client to understand the decision the client is trying to make and work with the client to help establish the decision rule. More and more laboratories are trying to provide analytical services that are designed to meet clients' needs, and the DQO process is one method you can use to establish what type and quality of data needs to be generated. The data quality objectives in turn become part of the QAPjP.

Please refer to the QAPjP in Appendix II of this book. It can be used as a generic QAPjP and a mechanism to generate DQOs for specific analytical testing projects.

Project Planning — Using the Checklist

Many laboratories have developed a checklist to help them scope out projects, tailor services to client needs, and estimate prices. These checklists are designed to help the customer and laboratory agree on deliverables, timing, prices, etc.; everything involved in the project. An example of a checklist is provided in Figure 1-2. This checklist provides the first cut at a project specific quality assurance plan (i.e., a QAPjP).

After you obtain the basic information regarding the name of the company, the contact person making the request, their address, phone number, and billing information (i.e., the internal project code or account number used by the company), you should get a brief description of the project or the site where the samples will be taken. For example, samples may be taken of suspect contaminated soil near a wood treating operation. From experience you might know that pentachlorophenol (PCP) will be an important analyte. You might also go over with the client whether there may be particular hazards associated with the samples. For example, metal treating wastes may contain percent levels of cyanide. In this case, you will need to make sure that sample digestions are conducted in well-ventilated hoods, etc.

You should get a clear idea of what turn-around time the client expects for this project. For most laboratories performing environmental testing you can expect normal turn-around times from about 45 to 60 days. Sometimes turn-around may be shorter depending on how busy the lab is. For mixed waste samples (i.e., samples that are radioactive), turn-around times will generally be longer — on the order of 60 to 90 days. Turn-around for mixed waste can be even longer, depending on the sample

FIGURE 1-2. The Checklist for Analytical Work.

Date ______________________________

Name of Originator ______________________________

Company Name ______________________________

Contact Person ______________________________

Alternate Contact in case of emergency ______________________________

Contract Name and Number ______________________________

Program and Project Name ______________________________

Project Description ______________________________

____

Do special Hazards exist ______________________________

____

Turnaround Time ______________________________

Available documentation (e.g. Contract, work plans, QAPjP, sampling plan, etc.)

____

Laboratory Certification required (list States) ______________________________

Samples being taken for regulatory compliance (RCRA, CERCLA, CAA, NPDES, NPDWA, etc.) ______________________________

Required Methods (SW-846, 600 series, 500 series, ASTM, Standard Methods, etc.)

____

Quality Assurance and Control ______________________________

- Samples and duplicates ______________________________
- Matrix spikes and duplicates ______________________________
- Methods blanks ______________________________
- Equipment rinsates ______________________________
- Trip blanks ______________________________
- Laboratory Control Samples ______________________________
- Sample volume for QC samples ______________________________
- Who will designate QC samples ______________________________
- Detection limit and reporting limit requirements (regulatory) ______________________________
- Additional QC ______________________________

Chain of Custody ______________________________

- Sent to laboratory ______________________________
- Send back with data ______________________________

Can we Subcontract analyses ______________________________

FIGURE 1-2. (continued)

Results

Hardcopy Reporting Format ____________________

Verbal ____________________

FAX ____________________

Diskette ____________________

Shipment method ____________________

____

____

Sampling schedule

When will samples arrive ____________________

Matrix ____________________

Sampling bottles needed ____________________

Samples

Storage requirements ____________________

Holding time after analysis ____________________

Return unused sample ____________________

Authorized representatives from the company for receipt of data

____

____

____

Holding time requirements ____________________

matrix and number of isotopes requested. Most labs offer rush service. This can be anywhere from 2 days up to the normal delivery. The important point here is to get an idea of what the client needs as far as turn-around time is concerned. Some labs may even be able to offer results within hours of sample receipt if the client needs the data quickly and is willing to pay for it.

Usually project managers do not think about laboratory services until the last minute. They generally assume that laboratory services will be available when they need them. Once a project is initiated the laboratory should try to get as much information about the project as possible. Ask to see the quality assurance project plan if it's available. If the work is being subcontracted to you, you might be able to get a copy of the contract so you can see what the work scope is. Many contracts, especially time and material types, use work plans for each sub-project. You might be able to get a copy of the work plan for the project if you ask for it. If available, you can also ask to see sampling and analysis plans or other standard operating procedures associated with the project.

You should establish whether the laboratory work needs to be performed by a state-certified laboratory. If the work is being done for clients in New Jersey, New York, Pennsylvania, Connecticut, Delaware, Maryland, or California, you may need to be participating in those states' Quality Assurance (Certification) programs.

You should find out what regulatory driver(s) is causing the work to be performed. Typically, the Resource Conservation and Recovery Act (RCRA), CERCLA (as modified by SARA), National Pollutant Discharge Elimination System (NPDES), NPDWA, CAA, and Environmental Compliance Agreements top the list of regulatory drivers encountered most frequently. You need to ascertain what target analytes will be of interest and whether the client wants you to use any specific methodology in performing the testing.

You need to find out what the client wants for quality assurance and quality control on this project. You might explain briefly what your standard QA procedures include. For example, blanks, batch QC, calibration, etc. Find out whether the standard QA that you use will be adequate for this project. If it's not adequate find out exactly what you need to do. You should discuss whether spikes, spike duplicates, and sample duplicates, etc., are needed. What they expect for the number of blanks (i.e., the blanks they will generate, such as trip blanks, field blanks, rinsates). You should explain how often you generate method blank data and whether they require a method blank to be associated with their specific batch of samples. You should explain that the client may need to take larger samples than normal if some samples will be used for QC purposes (i.e., to ensure there will be enough sample for spiking and reanalysis). You should find out if the client or you will be determining which samples will be used for QC purposes. Find out if there is any other QC the client needs that is not covered by the discussion above.

Since you may need to off-load some work to other laboratories you might find out if the client has any problems if you subcontract out some of the work. You should point out that if the need arises to use other laboratories you will use only laboratories that meet all of the requirements (i.e., certifications, QA, etc.) and the work will be performed at the negotiated price.

Most clients will want to have copies of the chain of custody forms and other documentation available at the end of the project. Make sure this is discussed. You also need to resolve whether there are holding time criteria outside of the normal holding times required by EPA. For example, when the client first calls the laboratory the samples may have already been taken and the holding time clock is already running. You need to find out if this is the case. You should also go over the standard methods you will use on the samples, their typical detection limits, and the fact that the detection limit is influenced by the matrix. Find out if there are any specific regulatory limits involved that will require detection limits below these limits.

The reporting format is another thing that you should discuss with the client. You will need to know whether they want/need a CLP deliverable or whether a Level II or Level III report will suffice. Different laboratories define Level II and III somewhat differently, as discussed above. The main difference in all of these reports is the level of QA/QC information provided, with the CLP report being the most extensive.

You also need to know whether the client wants the results verbally, followed with hardcopy; a facsimile, followed by hardcopy; a diskette (and the format for the diskette files — e.g., Lotus, DBase, CLP, etc.), or you may be able to send the file(s) via modem using one of several protocols. You also need to know whether the client wants the data via overnight shipment, mail, courier, etc.

One of the most important things to know is when the samples will arrive. Will they arrive in more than one container and will samples be taken and arriving over a period of time. You need to get as much of this information as possible to help you with scheduling work at your laboratory. You should verify how the samples will arrive. You should go over sample preservation and sampling containers with the client. You may need to provide sampling bottles and shipping coolers if the client needs them. You may also need to discuss sampling methods and preservation techniques to ensure that the sampling is done correctly and the samples are viable when they arrive at the lab. You should tell the client what your standard procedures are when you receive the samples. What you do when samples arrive without documentation, if sample bottles are broken, arrive late or out of holding times, etc. Most labs simply call the client to discuss what they want done when any out of the ordinary or unexpected event occurs.

You should also tell the client what your procedures are for unused sample. You may return it to the client or dispose of the material yourself. If you return samples you should make sure the client understands that unused samples will be returning and provisions need to be made to dispose of these samples. Often, you will need to store samples and extracts for some period of time until the data are accepted by the client. You should go over your storage policy (i.e., your policy may be to only store materials for 6 months, etc.).

Last, you also need to establish points of contact and authorized representatives from the client company. These contacts will receive the data and have authority to direct the laboratory to handle unusual conditions and to change previous terms or conditions. It might also be a good idea to have a designated contact for non-business hours in case of an emergency.

It is often important to know what decision will be made based on the data the laboratory provides. The lab should know whether the client wants to use the data to determine whether a waste is hazardous, or whether the data will support risk assessments, delisting petitions, license requirements, etc. The type and quality (and ultimately the cost) of the data the laboratory produces may vary depending on the use the data will be put to. In a sense, the laboratory should work with the customer to define data quality objectives.

In some cases it is important to have a low detection limit to support risk assessment decisions. In other cases the client may need data that span a wide concentration range. At other times the client may want assurances that false positives or negatives about a specific decision point are less than a certain amount (e.g., less than 5%). All of these will drive the laboratory's selection of method and ultimately dictate the cost of the project. Sometimes the laboratory will have little choice in the methods it must use because of specific regulatory requirements. For example, if samples are taken for drinking water evaluations then the lab will have to use EPA's 500 series methods. The Clean Water Act requires EPA's 600 series methods or ASTM methods for it's National Pollutant Discharge Elimination System (NPDES). The Resource Conservation and Recovery Act (RCRA) requires the use of SW-846 methods, and Superfund uses the IFB methods. Thus, regulatory requirements are significant when deciding on which method to use.

Sample and Reagent Disposal

Treating Hazardous Waste

Many laboratory managers believe they are not permitted to treat their hazardous waste. This belief is, however, incorrect. Obviously, on-site waste treatment has some real advantages for the laboratory. The best situation is to control how much hazardous waste you generate and thereby avoid disposal costs and potential liability.

What most laboratory managers don't know is that they have legitimate waste treatment options. Not all hazardous laboratory waste has to be disposed of as a hazardous waste. RCRA regulations permit certain types of treatment. You may treat RCRA characteristic wastes to render them nonhazardous by treating them in an accumulation container (40 CFR Part 265 subpart I) so long as you file a waste testing plan with your regulating authority. This exemption was explained in the Federal Register on March 24th, 1986 (51 FR 10168). Not all states recognize this exemption, however, so you need to check with your state to find out if you can apply this exemption. Before we get into your treatment options, let's review what wastes are allowed to be treated and permitting options available.

RCRA divides all of the wastes generated into two broad categories, hazardous and nonhazardous. Nonhazardous waste can be disposed of as ordinary trash and sent to the local landfill. A waste may be hazardous by exhibiting a characteristic of a hazardous waste or it may be a "listed" hazardous waste.

Characteristic wastes exhibit one or more hazardous characteristics (i.e., corrosive, ignitable, reactive, toxic). Laboratories will typically generate corrosive and toxic wastes and may also generate some reactive waste. Corrosive waste will be spent acid and base solutions. These wastes can be neutralized and disposed of down the drain. Many acid solutions may also be toxic due to high concentrations of RCRA metals (i.e., arsenic, barium, cadmium, chromium, lead, mercury, selenium, and silver). Normally, when you raise the pH of an acid solution containing metals, several metals will form hydroxides and precipitate out of solution. One of the best methods for treating acidic wastes containing metals is to raise the pH using a calcium hydroxide slurry. The pH of the slurry is 12.3 so you don't run the risk of having this material classified as a corrosive waste. Add the slurry while mixing the waste until you get to a pH of about 10. This will flocculate and precipitate most of the RCRA hazardous metals. Next add enough of a 2 *M* solution of sodium sulfide to get between 1 and 3 ppm of sulfide in solution. When the sulfide is added the pH of the solution will increase since the metal hydroxides will be converted to sulfides. If the pH rises above 12.5 you can bring it down by adding a little 0.1 *M* nitric acid.

This procedure will remove cadmium, chromium (II and III), mercury, lead, and silver.[3]

After this treatment, all you need to do is filter or decant the solution. The supernate can be disposed of down the drain. The filter cake can be disposed of as hazardous or you can mix the cake into a cement matrix and dispose of it as a nonhazardous waste. If you do dispose of it as hazardous your disposal costs will be much lower than disposing of the original acid solutions. This treatment can be performed without an RCRA Part B treatment permit as long as you have an approved waste analysis plan and treatment is performed in a closed container.

The laboratory also typically generates spent solvents. Methylene chloride is used in many extractions and can be recycled. After an extraction you will have a methylene chloride solution that contains semivolatile organic chemicals. The sample that was extracted will likely be nonhazardous when it is dry. Now you normally dry the extract with sodium sulfate. When the sodium sulfate is dry it also will likely be nonhazardous. The next step is to distill off about 99.5% of the methylene chloride from the extract. The methylene chloride can be condensed using a Buchi, passed through activated charcoal and reused. If you recycle methylene chloride you will have to distill an aliquot of it down and analyze it to show you got all of the impurities out. Thus, you will have very little methylene chloride to dispose of as a hazardous waste. You can do the same thing with hexane-acetone extracts. Treatment techniques have also been worked out for aqueous solutions with small amounts of dissolved solvents.[4]

The laboratory may generate small amounts of reactive waste (i.e., waste containing cyanide or sulfide). Cyanide can be removed by chlorination with hypochlorite and sulfides can be precipitated with copper chloride.

So you see, by filing a waste analysis plan, you can avoid generating most hazardous waste in your lab. All you need to do is show how you will test each batch of waste to prove it is nonhazardous prior to disposal. This should be relatively easy for the laboratory since the lab is already set up to test wastes. You can efficiently employ this approach by combining your acidic wastes and segregating your solvent wastes. You can accumulate acid wastes for a month or two and treat the waste in a batch. Solvents should be segregated to facilitate recovery (see Figure 1-3).

Listed Wastes — Some Advice

There are two types of "listed" hazardous wastes. Those that are commercial chemical products and waste streams from specific and nonspecific processes. For example, benzene is a toxic chemical. If you dispose of benzene then it must be disposed of as a hazardous waste. Let's suppose you have some benzene in the lab and, due to the change in methods and procedures in the lab you no longer use it as a solvent. You want to dispose of the benzene. Since the benzene is a commercial chemical product you have to dispose of it as a hazardous waste.

Compare that situation with the following. You have a waste which contains a small amount of benzene. Is it a listed hazardous waste because it contains benzene? Probably not because benzene is listed as hazardous when it is a commercial

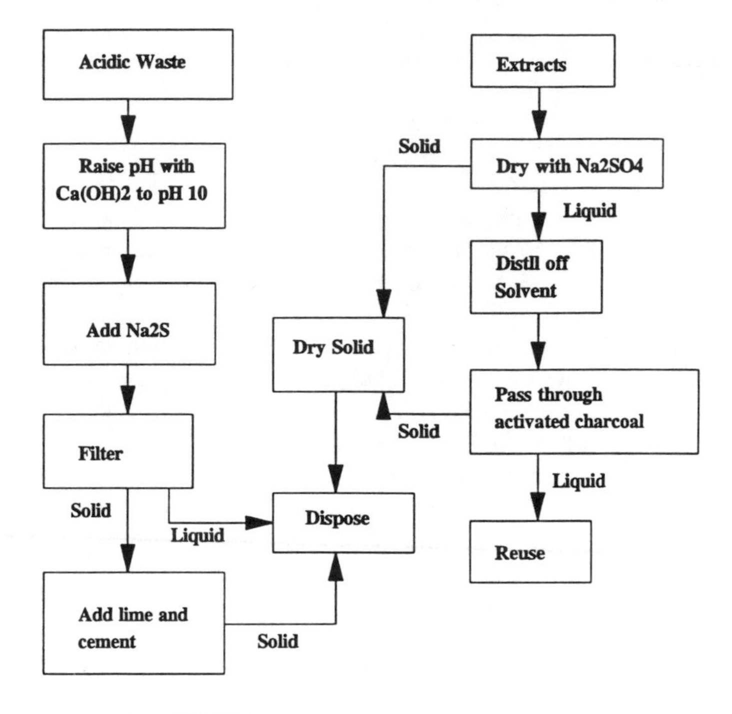

FIGURE 1-3. Waste treatment scheme.

chemical. So in this case the waste may be toxic but it is not a listed waste. This is an important difference since you cannot treat listed wastes without a permit but you can treat characteristics wastes. Lets take this example further. Suppose there are two wastes. They are identical in that they contain benzene. One waste is characteristic and contains 1% benzene. The other waste is listed and also contains 1% benzene. In the lab you may remove the benzene from the characteristic waste without a permit and dispose of the waste as nonhazardous (you have to have a waste analysis plan to do this). In the case of the listed waste, however, you cannot treat the waste unless you have a full Part B Permit. Most labs will not have a Part B permit since it is time consuming and expensive. Therefore, you will have to dispose of the listed waste as hazardous. Same waste — two different disposals. One costs about 10 times more than the other. The point is you have to know what type of waste you are testing and that, in turn, will determine what options you can legally pursue for disposal.

A Final Bit of Advice

After observing several laboratory managers and their management styles, I've concluded that the laboratory manager should spend at least 50% of the time in the laboratory. The manager should wander around, ask questions of the analysts and technicians, ask how the work is going, ask if there is anything that might make the process more efficient. Managers that stay involved in the day-to-day operations of the laboratory are, in my opinion, more likely to recognize problems and take appropriate action before problems become serious.

REFERENCES

1. **Taylor, J. H. and Amano, R. M.,** *J. Chromatograph. Sci.*, 25 (8), 364, 1987.
2. **Martin, M. J.,** *Environ. Lab,* 5(3), 14, 1993.
3. **Goronszy, M. C., Eckenfelder, W. W., and Froelich, E.,** A guide to industrial pretreatment: waste water, a variety of mechanical and biological methods are at your disposal, *Chem. Eng.,* 78–83, 1992.
4. **Grulich, G. Hutton, D. G., Robertaccio, F. L., and Glotzer, H. L.,** Treatment of organic chemicals plant wastewater with DuPont PACT process, *AIChE Sym Ser.*, 129(69), 127–133, 1972.

CHAPTER 2

The Metals Laboratory

The first portion of this chapter provides general information as background. The later portion provides practical instructions and techniques for performing metals determinations. In most laboratories metals analysis is performed by a metals group (see Figure 1-1). This group is responsible for aliquoting the sample, digesting it (or otherwise preparing it for analysis), sample analysis, and data reporting, including quality control.

In this section we examine how samples are prepared and determined and the choices one can make regarding these factors. We also include information regarding the way samples should be prepared and analyzed.

GENERAL INFORMATION — MATRICES

Typical matrices include water, soils, sludge, organics, and air filters. Depending on the project, samples may include plants, animals, and biota. Samples may also be multiphasic.

Water

Water samples are fairly common in most laboratories and generally represent the easiest matrix to deal with. Water comes in several varieties and it is important to understand the differences. There are surface, drinking, ground, and waste waters.

Water quality varies and some samples of groundwater or wastewater will have very high dissolved solids (10,000 ppm or more) while others will be low. The high solids content will likely produce interferences for some metals. All water samples may contain sediment. It is often difficult to interpret the results from these types of samples since they are normally acidified in the field prior to shipment to the laboratory. Thus, the sediment is leached by the acid that has been added to the sample. Aqueous samples may also come from a leaching procedure. Aqueous extracts from leaching tests will usually contain acetic acid buffer or a dilute mixture of sulfuric and nitric acids.

There are many ways to analyze water samples. You can analyze for total dissolved metals, or total metals, or total available metals. Each of these analyses involves a different preparation procedure and you will generally get different results depending on the preparation method used.

Soil

Another common matrix is soil. Soil samples may look homogeneous but rarely are. Results from successive 1-g soil samples may vary by 100% or more for some elements. There is also the problem of debris mixed in with the soil. In a case like this you can grind the soil sample (including the debris) to get a homogeneous sample and then remove sub-samples for analysis. Another way around this problem is to sieve out foreign material (i.e., leaves, stones, twigs, etc.) and mix the soil to get as homogeneous a sample as possible. If you analyze samples in replicate, the relative percent difference or standard deviation will probably reflect how well the soil was sieved and mixed, not the precision of your analytical technique.

Lead abatement is becoming an important remedial activity in urban areas. Soils in older communities are often heavily contaminated with lead from leaded house paint. These soils contain small chips of paint that are indistinguishable from the soil. On the addition of acid, however, you might notice that small chips float to the top of the acid where they effervesce. The effervescence is due to the release of carbon dioxide from the lead carbonate, which was a popular pigment in white house paint before 1960. The lead concentration will be highest in the soil samples taken closest to the house and in areas where rainwater drains away from painted surfaces. We have seen soils that contain up to 5000 ppm of lead due to leaded house paints.

Sludges (Metal Hydroxides)

Metal hydroxide sludges typically result from treatment of wastewater from metal finishing operations. When the pH of the wastewater is adjusted (usually the pH is increased by adding lime or caustic) many metals will form insoluble hydroxides which precipitate from solution. Floculants might also be added to aid in the precipitation. The sludge is dewatered by filtration and the wet filter cake is disposed of. The treated wastewater is either sent to a sewer or discharged to surface water under a National Pollutant Discharge Elimination System (NPDES) Permit. The resulting sludge can present several analytical challenges since it will have high concentrations of some metals (e.g., iron) that may interfere with analytes of interest. Digestion of these sludges is similar to soils except you may need to add more acid to overcome the effect of the lime or caustic.

Organics

Laboratories are often faced with analysis of organic matrices for metals. The organic matrix may be a solvent, an oil, or some other organic or polymeric material. These samples are generally the most difficult to analyze depending on the metals of interest. In this case, sample preparation can be difficult and dangerous if not performed properly.

Oils and other organic material can be digested with acids or a mixture of acid and peroxide. If organics must be digested, the best approach is to use either a bomb technique or microwave digestion. Either of these approaches will rapidly destroy the organic material and leave a solution that may be aspirated into the atomic absorption (AA), or inductively coupled plasma (ICP) (after dilution — see effect on detection limits below). Oils may also be diluted with solvent and aspirated directly.

Organic solvents are not readily digested since they evaporate in an open beaker or build up high pressure and possibly explode in a bomb or microwave digestion vessel when heated. There are three basic ways you can analyze solvents for metals. The first option is to remove the metals from the solvent by washing the solvent with an acid digestion mixture. The solvent is placed in a separatory funnel and washed with a mixture of nitric (2 to 3%) and hydrochloric acids (0.5 to 1%). Metal particulates in the solvent will be dissolved by the acid. Organometallic compounds are readily destroyed by acid and the metals will be extracted into the aqueous acid phase. This technique is commonly used for the determination of lead in gasoline since the tetraethylead (actually a mixture of methyl and ethyl leads) is destroyed by the acid.

The second option is to allow the solvent to evaporate slowly and then digest the residue. We have used this approach by letting aliquots of the sample evaporate under a hood overnight (covered loosely to prevent foreign material from falling into the sample). Spike recoveries were very good (above 90%); however, spikes were made with organometallic standards that had a lower volatility than the solvent.

The third option is to dilute the sample with another solvent and aspirate the diluted sample directly into the ICP or flame AA. In this case, standard additions or an internal standard should be used since the viscosity of the standards and the unknowns will be different.

Air

Metals in air are determined by drawing a known volume of air through a high-efficiency particulate air (HEPA) filter. Filters are precleaned with acid to remove any trace metals prior to use. After the air sample is drawn through the filter it is leached with hot acid and sonicated to dissolve the metals prior to analysis. Metal recovery is generally very good (near 100%) as determined by spiking with soluble metal salts. Even though spike recoveries are good, we know that some metals are not recovered very well (e.g., chromium). This is due to the state of the metal on the filter. There are some insoluble metal compounds in air particulates that are not dissolved by hot acid leaching.

TIPS AND TECHNIQUES — SAMPLE PREPARATION

There are literally hundreds of sample preparation methods available that are optimal for just about any matrix and metal combination you're likely to come across. Unfortunately, with environmental samples, we are often asked to determine 20 to 30 elements and there is no single best digestion method for all elements in all matrices. Therefore, in the interest of speed and safety we sacrifice accuracy and, to some extent, precision, when we digest environmental samples.

The most common digestion methods rely on hot acid leaching; typically nitric or a mixture of nitric and hydrochloric acids are used for the determination of 20 to 30 elements. Some elements (materials), however, cannot be dissolved in this way. Silica, for example may dissolve initially; however, on standing it may polymerize and fall out of solution. The sample preparation methods we discuss here will not generally work for silica since silica is normally not of interest. This is not a problem for most environmental laboratories. If, however, you literally want to analyze samples for total metals you will have to totally digest the sample including silicates or other insoluble species. Analysis of certain matrices for radioactivity often requires total dissolution of the sample. This is the situation when the matrix was subjected to high temperatures or pressure.

The methods provided in this section are general-purpose multielement methods that are the most common methods used by environmental analytical laboratories.

Sample Size and Detection Limit

Detection Limit depends on sample size

If the sample will be digested prior to analysis (i.e., if you are analyzing for anything other than dissolved metals), typically, 100 ml is used for aqueous samples, about 1 g for soils or other solids, and 0.25 to 0.5 g if the sample is organic. These sample sizes may be increased or decreased depending on the required detection limit.

Detection limits for aqueous samples when using ICP-AES will generally be in the 50 ppb range or somewhat lower. Furnace AA determinations can often be performed at about 1 ppb. ICP-MS is the most sensitive of the techniques and can be used to determine a wide range of elements on the order of 0.1 ppb. Detection limits for soils are typically higher than aqueous samples by a factor of about 100 due to the smaller sample size and increased dilution. Samples larger than a few grams become too cumbersome to handle conveniently, thus limiting the detection limit. Typical detection limits for soils (ICP-AES) are around 0.5 ppm. Detection limits for oils and other organic matrices are higher still (10 ppm or higher) due to the small sample size and dilution. This problem may be overcome, to some extent, by using the direct aspiration method (see below).

Acid Selection and Acid Strength

For most digestions either nitric acid or a mixture of nitric and hydrochloric acid is used. The acid(s) that is used will, to some extent, determine the analytical technique that may be used for the metal determination. High chloride concentration will cause problems if the sample will be analyzed by ICP/MS. Therefore, only small amounts of HCl are used (or it is eliminated altogether) when performing digestions for ICP/MS determination. Furnace determinations typically use only nitric acid while flame and ICP determinations use mixed nitric and hydrochloric acids. This is explained below.

The purpose of the digestion is to get the elements of interest into a form (solution) that is convenient to analyze. Therefore, most digestions use an oxidizing acid (nitric, sulfuric, or perchloric) and potentially other oxidizing agents (i.e., hydrogen peroxide) to insure metals are in solution.

Digestions for AA, ICP-AES, and ICP/MS

Digestion is recommended for aqueous samples even if they are clear

Water samples may be analyzed by simply filtering through a 0.45-µm filter that has been prewashed with a little 0.5% nitric acid. This will give a concentration of the *total dissolved metals* in the aqueous sample. Depending on the water matrix, however, it may be necessary to digest the water sample by heating with nitric and hydrochloric acids. There are only a few documented cases where the digestion of a clear aqueous sample improved metal recovery. However, since it is not possible to predict *a priori* when this will be the case, the safest approach is to digest all aqueous samples. The client's needs or specific regulation will ultimately determine how you prepare the sample and the analytical technique you must use.

After the samples are digested they are either decanted or filtered (0.45 mm) to remove insoluble material. Filters should be washed with a little dilute nitric acid to remove any metals that may be present before filtering the samples. It is very common when digesting some waters and soils that silicates will precipitate out of the digestion solution. These insoluble materials are filtered out using the 0.45-mm filter. The filtrate is placed in a polyethylene bottle and is stored at room temperature until analysis. The holding time for metals analysis is normally 180 days after digestion, with the exception of mercury.

Precipitates may form during storage in samples containing high concentrations of dissolved solids. If this occurs you can try to redissolve the precipitate by warming, adding additional acid, or diluting the sample. These precipitates are sometimes difficult to redissolve. If you cannot redissolve the precipitate, the best approach is to get a fresh (smaller) aliquot of sample and digest and analyze quickly. Alternately you could filter the sample and analyze the supernate solution if you cannot redissolve the precipitate. This condition is noted on the sample report.

Digestions for Furnace AA

Nitric acid alone or nitric acid followed by peroxide is typically used for furnace work. Until recently, mixed acid digestions (i.e., nitric and hydrochloric acids) were not recommended for furnace work because of the volatility differences between chlorides and nitrates for some elements. That is, some of the sample may volatilize at one temperature (corresponding to the chloride) while the rest of the sample would volatilize at another temperature (corresponding to the nitrate). Some metal chlorides are relatively volatile and may be lost during the char stage. This situation has changed; with the advent of pyrolytic graphite tubes and platforms mixed acid digestions can be used for graphite furnace AA. This simplifies the digestion procedure for the analyst since one digestion method can be used for ICP-AES, FAA, GFAA, and ICP/MS.

With the advent of improved graphite tube design, such as the L'vov platform, the sample is vaporized by radiant heat rather than convection, leading to a slower release of atoms. In this case, the atoms are vaporized from the platform rather than from the walls of the graphite tube. Because of this slower release, the sample is atomized during the measurement phase of the determination and there is much less of an effect from mixed salts. Furthermore, because the temperature within the tube is higher when the sample is vaporized, chemical interferences (i.e., in the form of the recombination of atoms) are less likely.

Heating Methods

When performing a hot acid digestion, regardless of the matrix, there are several methods of heating the sample. Each one has its own advantages and can be used in different situations.

Hot Plate

The hot plate is the most common and least expensive sample digestion method. For water samples one typically starts with 100 ml of aqueous solution in a beaker (250 ml or larger) and adds 3 to 5 ml of nitric acid depending on the specific procedure. This solution is put on the hot plate and heated to just below its boiling point (about 85 to 90°C) with gentle refluxing along the sides of the beaker. The sample volume is reduced by a specified amount, typically from 15 to 90%. After evaporating the prescribed amount of liquid, the beaker is cooled slightly and concentrated hydrochloric acid is added (1 to 3 ml). The addition of the HCl will dissolve most of any precipitate that has formed with the exception of silicates. If a precipitate is present the sample may be decanted or the insoluble material filtered off at this time. The digestate is taken up to the original volume (100 ml) and analyzed directly

Ribbed watch glasses speed up evaporation during reflux

Most methods call for covering the sample with a watch glass during the digestion. The cover is useful to aid in the reflux of the sample and acid and to keep foreign material from getting into the beaker. If you use a regular watch glass (i.e., the kind without the ribs) to cover the beaker the evaporation takes two to three times longer. The ribbed watch glasses make more suitable covers for acid digestions. Until recently, these ribbed watch glasses were not available. Corning started to make ribbed watch glasses a little over a year ago and they come in three sizes; 75, 100, and 125 mm. The Corning part numbers are 9990-75, 9990-100, and 9990-125, respectively. Use of ribbed watch glasses saves significant amounts of time if you're using digestion procedures that call for sample evaporation. You can now order these ribbed watch glasses from standard glassware suppliers (i.e., Fisher, VWR, etc.). Because of the unavailability of ribbed watch glasses and the need for speed, some labs don't cover the beaker at all to enhance evaporation. We have not tried open beaker digestions. It probably works fine for most metals; however, you might lose some of the more volatile elements and introduce contaminants. Since ribbed watch glasses are now available, it's a good idea to keep the beaker covered.

The problem with hot plate digestion is if you go too fast (i.e., boil the solution with or without a watch glass or let the sample go to dryness) you may lose some of the more volatile metals such as selenium, arsenic, and possibly lead, and recoveries will be low. Since the hot plate is, more or less, an open system, sample loss can occur.

Depending on where the beaker is on the hot plate its temperature will vary. Some samples will get too hot and may boil, while others take longer to reduce in volume. One way around this is to keep the temperature of the hot plate as low as practical (i.e., so the temperature in the beaker doesn't get above 85°C) and to move the beakers around on the plate. Some labs use a heat distributor, such as a metal block on top of the hot plate to maintain as even a temperature as possible. If a sample boils a little it will not normally affect the results. Monitoring spike recoveries is about the only way to know whether your digestion technique is adequate or whether you are losing some elements. You can also use the results from laboratory control samples (LCS) or standard reference material (SRM). You should plot the recoveries from the LCS and spikes to establish a control chart for the procedure you are using. The control chart will show how well the procedure works in your laboratory over time.

When digesting solid samples you essentially follow the same procedure as aqueous samples except you use a much smaller sample and perhaps a slightly different acid mixture. In this case the solid (normally between 1 and 2 g) is leached with hot nitric acid. After hot acid leaching, concentrated hydrogen peroxide is added to oxidize any organic matter. If there is more than a percent or two of organics the peroxide addition should be made very slowly, since peroxide will react violently with organic material. The reaction is continued under gentle reflux until the digestate is clear and lightly colored. Normally a slight yellow or straw color is a good end point. Since this is a hot plate digestion you still have the same problem of boiling and evaporation as mentioned above. When the digestion is finished the digestate is diluted up to a convenient volume. The dilution increases the detection limit by that amount. As indicated earlier, solid sample larger than a few grams is inconvenient. Thus, you are limited to a 1-g sample due to practical considerations and this, in turn, increases detection limits. For example, if a 1-g sample is digested and the final volume is 100 ml, the detection limit will be 100 times higher than for a water sample, assuming everything else is equal. Thus, the detection limits for solids are always substantially higher than for liquids.

Multiple phases should be analyzed separately

If a sample has more than one phase you can either try to homogenize the sample and take an aliquot or analyze the phases separately and report the result for each phase. Often, it is very difficult to homogenize samples with liquid and solid phases or samples with organic and aqueous phases. The problem with mixed-phase samples is they begin to separate while you're trying to get an aliquot. Thus, an aliquot is likely to be biased since you will get more of one phase than another. We have found that the best approach is to analyze each phase separately and either report the results from each phase or report a mathematically weighted average based on the weights of each phase. This is important if you are trying to determine whether a sample is a hazardous waste under the Resource Conservation and Recovery Act (RCRA). For example, if a sample is 20% solid, 50% aqueous liquid, and 30% organic liquid, by weight, then the result would be 0.2 of the solid concentration, 0.5 of the aqueous liquid, and 0.3 of the organic. Detection limits are handled in the same way. If the sample has undetectable concentrations of lead and the detection limits were 0.5, 0.05, and 5 ppm, respectively, for the three phases, then the reported detection limit would be 1.6 ppm.

Microwave

Microwave-assisted digestion is a relatively new and convenient digestion method. Up to 12 samples can be digested at one time. The number of samples that can be digested depends on the power characteristics of the oven you are using. For water samples you add 50 ml of the sample along with nitric acid and set the power levels on the microwave oven to reach the specified temperature (and pressure) in the liquid and hold that temperature for a specific time. This is covered in proposed methods 3051 and 3015 in EPA's Methods Manual SW-846.

Available microwave power dictates the number of samples you can analyze at one time

The total cycle time for up to 12 samples is about 1 hour when performing microwave digestions. The trick in using microwave digestion is to keep the power levels matched to the number of samples being analyzed. This insures uniform digestion conditions. That is, if the conditions in a method call for six samples of 50 ml and a power level of 100% (600 W) for 10 min, then you cannot add more samples to be digested unless you get a more powerful microwave oven. If you attempt to add six more samples and run the microwave at 100% power for 10 min all of the samples will absorb only one half of the power as before and the digestion temperatures will be different. The key to using microwave digestion reproducably is to make sure that the power conditions are the same for all samples. This will insure uniform heating and pressure conditions for the samples and thus reproducible results.

Microwave digestion is easy to use and almost foolproof. There are a few cases where one has to be careful, however. Never attempt to digest an organic solvent in a microwave digestion vessel. The pressure in the vessel will rise too fast and the vessel may explode. Samples that give off large volumes of gas need to be handled carefully if microwave digestion is to work properly. The digestion vessels are designed to vent and relieve some of the pressure if the internal pressure becomes too high (see Figure 2-1). When the vessel vents you may lose some of the elements that may be in the gas phase. If you attempt to digest volatile samples, the pressure inside the vessel might become too high before the vent device can compensate. In this case the vessel can explode.

Only small samples (0.25 to 0.5 g) of organic matrices and samples that contain carbonates should be attempted. The problem is that the more gas that is given off the higher the pressure in the vessel and the more likely the vessel will vent. Oil samples, for example, can be digested with 10 ml of nitric acid to 0.25 g of oil. Here, as with solid samples, the detection limit will be higher than with water samples because of the dilution. In this case the final digestate is taken to 50 ml with deionized water. This solution is 20% acid and may be difficult to aspirate into the ICP. Likewise, because of the viscosity, it is difficult to pipet into the graphite tube of the AA furnace.

Microwave-assisted digestion provides about the same recoveries as conventional hot plates. In one study using the SRM 2764 Buffalo River Sediment, recoveries for most elements were the same between the hot plate digestion and the microwave. The microwave digestions, however, provide much better precision than the hot plate. Microwave digestion is more expedient than conventional hot plate digestion since the microwave energy is concentrated in the digestion solution and, therefore, more energy is devoted to digestion. That is, microwave energy input is more efficient at getting power into the sample than the hot plate. Microwave digestion also has the advantage since it may be performed unattended, and is applicable to more matrices

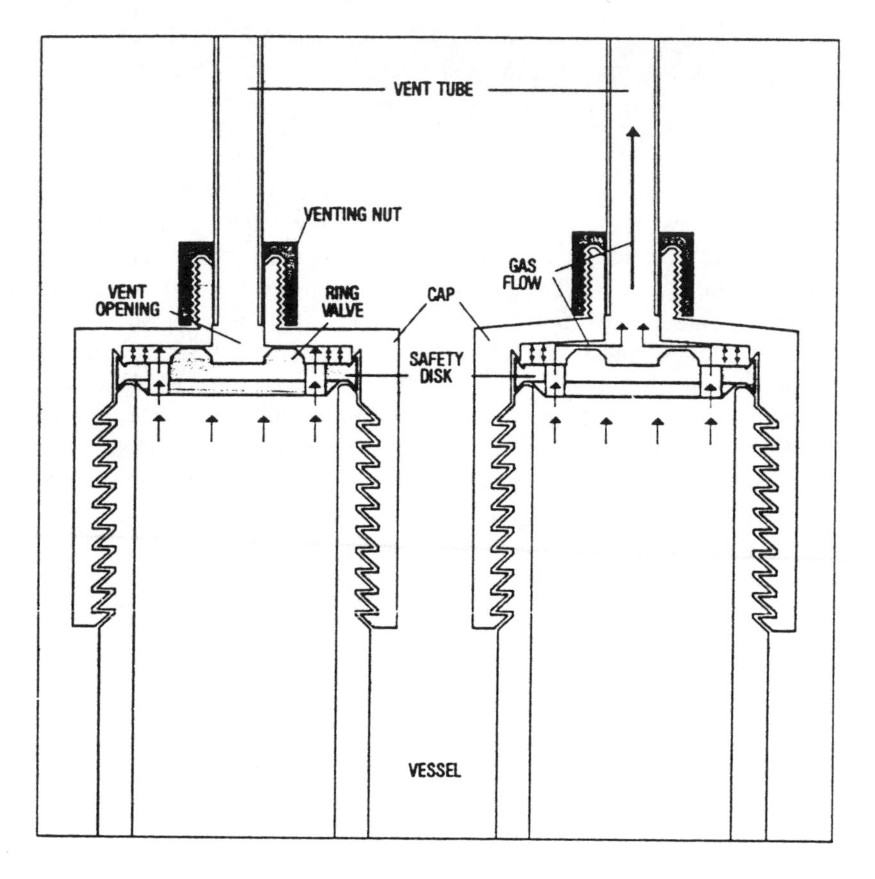

FIGURE 2-1. Diagram of microwave digestion vessel.

(i.e., oils and organic matrices) than the hot plate. Table 2-1 provides a comparison of recoveries between microwave and hot plate digestions for the Buffalo River Sediment.

Bombs

There are several types of bombs in use in the environmental laboratory. Some bombs are designed to oxidize the sample in an oxygen atmosphere while others simply use high pressure and high temperature. The latter type produces results similar to microwave digestion because the sample and acid are heated in a closed pressurized vessel. Oxygen bombs are commonly used for determining metals in organic matrices for the energy industry; however, they are not widely used by environmental laboratories. Because of this, we will only discuss the high pressure/temperature acid bombs in this section.

Bombs consist of an outer steel jacket and an inner crucible that is usually made of Teflon (see Figure 2-2). This type of bomb is very popular and may be purchased for around $150. For most digestions bomb temperatures may be between 150 and

TABLE 2-1. Comparison of Methods 3051 and 3050 for ICP Analysis of SRM 2764 Buffalo River Sediment (μg/g)

	Microwave			Hot Plate			%
Element	Mean	N	%RSD	Mean	N	%RSD	DIFF
Al %	1.18 ± 0.137	36	12	1.31 ± 0.367	24	28	–10
B	34.6 ± 9.31	30	27	55.4 ± 26.8	27	46	–38
Ba	77.7 ± 5.9	30	8	85.3 ± 17.9	27	21	–9
Be	0.562 ± 0.068	30	12	0.662 ± 0.209	25	31	–18
Ca %	2.00 ± 0.383	30	19	1.83 ± 0.200	27	11	+9
Cd	3.19 ± 0.613	29	19	3.32 ± 0.436	27	13	–4
Co	10.7 ± 1.46	30	14	11.1 ± 2.75	27	25	–4
Cr	81.7 ± 5.33	30	7	83.3 ± 14.0	27	17	–2
Cu	80.3 ± 6.92	30	9	83.2 ± 11.0	24	13	–3
Fe %	2.96 ± 0.214	30	7	3.06 ± 0.368	27	10	–3
Mg %	0.816 ± 0.047	30	6	0.850 ± 0.120	27	14	–5
Mn	460 ± 25.7	30	6	472 ± 57.2	27	12	–2
Ni	36.4 ± 2.52	27	7	37.7 ± 5.15	27	14	–3
Pb	143 ± 9.46	30	7	147 ± 16.0	27	11	–3
Sr	33.0 ± 2.05	30	6	35.0 ± 7.04	26	20	–6
V	21.0 ± 2.46	30	12	24.2 ± 7.21	25	30	–13
Zn	383 ± 26.5	30	7	393 ± 60.7	27	15	–2

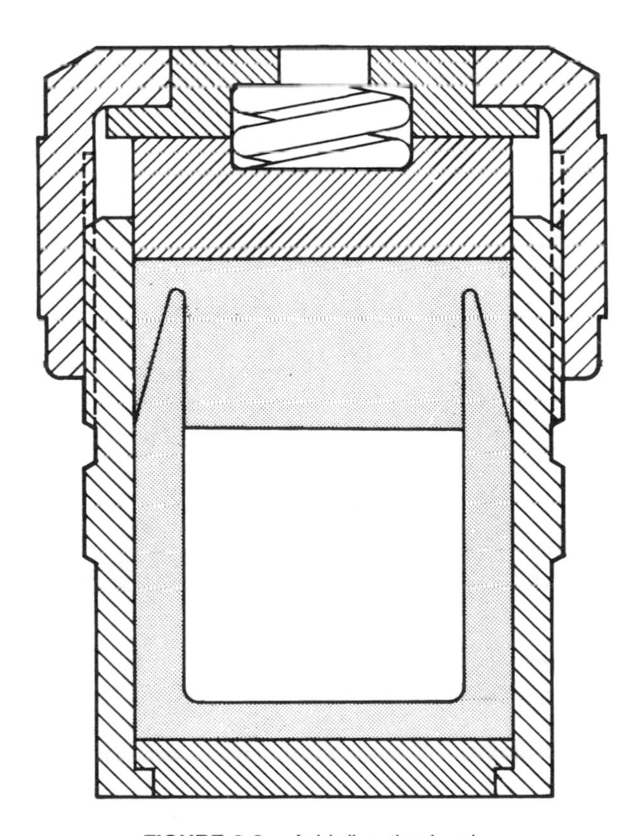

FIGURE 2-2. Acid digestion bomb.

275°C. Depending on the bomb, you may be limited to temperatures below 150°C because the Teflon in the crucible will creep and the bomb will leak. Bombs generally work well and they give comparable results to microwave digestion. As with microwave digestion, bomb techniques require small sample sizes and thus, dilution of the sample with acid increases the detection limit. This is further abrogated by addition of water to make the final acid strength 20% or less.

Our experience with bombs has been generally good except that some of them may leak, especially if you exceed 150°C. A small amount of the acid may leak from around the crucible and you may notice some dried, discolored material on the inside of the metal jacket and on the explosion disk. If leakage occurs it is usually due to a deformed Teflon crucible. Crucibles that leak should be replaced.

The procedure is to weigh the sample in the crucible, add acid, and place the covered crucible into the steel jacket. Some labs refrigerate the bombs so tightening is easier. The bomb is tightened and then heated at 150°C for several hours depending on the method being followed. We find it convenient to leave them in the oven overnight. If you refrigerate the bomb before tightening you might want to consider refrigeration before opening as well to lower the pressure.

We have found that the safest sample size for bomb digestion is about 0.25 g. The problem with this is the sample size is very small, which leads to higher detection limits and the possibility of sampling (sub-sampling) errors increases. If the sample is not homogeneous and you only use 0.25 g, you are likely to have significant sampling (subsampling) errors. The bombs we have used have about a 25-ml capacity, so they are limited to small samples.

Sample Dissolution for Organic Matrices

One of the best ways to handle an organic matrix is simply to dissolve a portion of the sample in a suitable solvent and aspirate the solvent directly into the flame or plasma. A known amount of sample (oil or grease) is dissolved in either kerosine or toluene and the solution is aspirated into the AA flame or ICP torch. We have successfully employed this technique by using a high solids nebulizer on the ICP and adding an internal standard to correct for matrix effects.

Sample dilution will be 5 to 1 minimum. Higher dilutions are desirable as long as you meet detection limit requirements.

When using this technique you have to make sure that the final sample is not too viscous to be aspirated into the flame or plasma. If the solution is too viscous, the delivery into the flame or torch will not be efficient. The sample must be diluted with kerosine or toluene (or other solvent) at a ratio greater than 5 to 1 to avoid aspiration problems. The greater the dilution ratio the more closely the sample viscosity behaves like the solvent and the less the need for an internal standard. We have used this technique at dilution factors as low as 5 to 1; however, at this dilution you will need to use an internal standard or standard additions to compensate for the difference in viscosity between samples and standards. The key factor one must control is the viscosity of the sample. If samples are run at low dilutions you will almost certainly have to use an internal standard (or standard additions) to get accurate results.

If you are going to run samples via ICP use of an internal standard is recommended. The internal standard makes it easier to take matrix effects (viscosity) into account. The internal standard must obviously be an element that is not present in the sample. We have successfully used scandium and yttrium as internal standards with oil and grease samples. If you are using flame AA then the method of standard additions (MSA) is recommended since MSA is the only way to get an accurate result using a single element technique.

When using dissolution methods you should prepare standards in the same or similar matrix. This means that the standards must be organometallic compounds. CONOSTAN™, a division of CONOCO in Ponca City, OK, can supply all of the standards, blank oil, metal stabilizers, etc., that you will need to use this sample preparation method. We use kerosine from Fisher Scientific as our solvent for oil analyses. When you make up solutions in kerosine you need to add metal stabilizers to ensure the stability of the organometallic standards.

We have found that the method works best at the highest dilution you can tolerate and still get required detection limits. That is, the higher the ratio of solvent to sample the better, since the viscosity of the solvent is low. For example, if you want to calibrate down to 1 ppm and you can accurately determine a standard at 0.1 ppm, you can dilute the sample 9 to 1 in kerosine.

As we have indicated, the sample dissolution method works quite well for metals in oil or organic matrices. This method, however, does not give total metals when there are particulates in the sample. If the sample contains metal particles, soil, sediment, etc., this material will end up at the bottom of the sample container after dilution. Therefore, this analysis will not represent "total" metals if there are appreciable amounts of solid materials present in the sample. If this is the case, and you want "total" metals, you should use either the microwave or bomb digestion techniques.

Use of Blanks and Quality Control Sample Information

Blanks and quality control QC samples are normally required for each batch of samples. A batch is a group of samples that is processed at the same time using the same procedure. Blanks and spikes are used to monitor laboratory conditions. A high blank will indicate that you may have a reagent, glassware, or other house-keeping problem.

Fortified blanks (i.e., a blank spiked with analytes of interest) are used to determine whether the analysis for a batch of samples was conducted correctly. This type of sample can also serve as a laboratory control sample (LCS). If the results for the fortified sample are within acceptable limits (typically within two to three sigma of the average value) then the analysis procedure is behaving normally and the analysis method is in control. One sample from each batch is usually selected for spiking, either singly or in duplicate. The spiked sample provides information about how the analytical procedure is performing in that matrix. Good recovery of spiked metals (i.e., greater than 80%) is desirable in order to conclude that sample analysis is accurate.

Special Cases — Mercury and Silver

Mercury

The standard method for determining mercury is to digest the sample with permanganate. After digestion the mercury is in solution as mercury ion, which is then reduced to free mercury with tin. The mercury is then determined by atomic absorption with a mercury lamp. This procedure is very sensitive, which can cause problems if samples have too high a mercury concentration.

This technique is sensitive to mercury in the sub-parts-per-billion range because the mercury in the entire sample is determined at one time. For example, if the sample size is 50 ml, then all of the mercury in the entire 50 ml is purged out as free mercury and determined as a "cold vapor". The sample size in this case is literally hundreds of times larger than the sample size aspirated into the ICP or AA. Thus, the detection limit for mercury is several orders of magnitude lower than other metals. In fact, any determination that can be made using a larger sample, as in the case of metals as their gaseous hydrides or cold vapor, will be much more sensitive than conventional ICP or AA.

This increase in sensitivity is useful when you need low detection limits. However, when analyzing solid wastes that have appreciable concentrations of mercury, this technique can be troublesome. High concentrations of mercury from one sample can contaminate plastic tubing, sparging glassware, and sample (digestion) bottles. Cleaning mercury out of a contaminated system is time consuming. Sometimes it's more cost effective to throw the contaminated glassware away and get new glassware.

We measured a sample that contained several hundred ppm of mercury. That sample seriously contaminated the cold vapor system. We subjected the glassware to several washings with hot permanganate, baked it, and swept gas over it for several days in order to get the background down to previous levels. We had to replace all of the tubing in the system. This was one of a series of samples we had to analyze for mercury and we lost several days due to this problem and learned that if you have any doubt about concentration you should dilute the sample and analyze it before analyzing undiluted digests. Although we were eventually able to analyze the samples within holding times (the current holding time for samples for mercury determination is 28 days because of its volatility) it was a close call and could have been an expensive mistake.

Silver

Silver forms a precipitate with chloride and, therefore, hydrochloric acid is not normally recommended as part of the digestion mixture. Silver, however, can be determined successfully using the standard nitric and hydrochloric acid mixture as long as the silver concentration in the sample is less than about 2 ppm. Therefore, as long as a sample contains less than 2 ppm silver no special digestion for silver is required. Unfortunately, simple salt (NaCl) is ubiquitous in the environment and most samples contain appreciable concentrations of chloride. Thus, silver recovery for many samples spiked with silver will be low because of the additional chloride.

METALS DETERMINATIONS

Most of the problems we have experienced with metals determination are more sample related than instrumental. Current instruments are relatively simple and fairly reliable so most of the problems you will encounter are associated with sample matrix. Matrix effects are relatively common when making metals determinations. The effect from the matrix is due to the difference in viscosity between the standard and the sample. When you aspirate standards that have a low viscosity you deliver the maximum amount of sample to the flame or torch. When you aspirate a sample with a higher viscosity, less sample is delivered and potential errors arise. You can tell if you have a matrix effect by diluting the sample and re-analyzing. If the two results agree (taking the dilution factor into account) then a matrix effect is not present. This means that the viscosity of the sample, after dilution, is about the same as before the dilution, hence no effect. If the measured concentration after dilution differs by more than about 20% then a matrix effect is suspected. One common way to compensate for a matrix effect is to use standard additions.

Standard Additions

By definition, the viscosity of the sample solution will be different from your standard. The problem is that if the viscosity of the sample is *too* different you get the wrong answer and you need to perform standard additions. The difference in viscosity also means that the slope of the standard additions line (i.e., the one you construct using the sample solution) will be different from the slope of the standard calibration line. The slope of the standard addition line tells you that for a certain increase in concentration you get a specific increase in signal.

Standard additions can be used to correct for matrix effects in ICP and AA techniques. We have all seen the diagrams of how you spike the sample and re-analyze it and plot the results. I've never really understood that diagram very well. The way I think about standard additions is that it's just the same as making a calibration curve, except you use the sample as the diluent. You literally construct a calibration curve using the sample solution. Normally, when you make a calibration curve you use deionized water to dilute a concentrated standard. When you perform standard additions you take three aliquots of the sample solution and spike in a small volume of a concentrated standard to add say 0.5, 2, and 5 times the concentration of the element you are trying to determine.

Let's say you measure a sample and, from the calibration curve you estimate that the concentration of the sample is 1 ppm. Let's call this absorbance 1 unit. Now you adjust the concentration in the sample by adding enough standard to make the concentration 2 ppm. If there is no sample transport effect you should get a signal that is equivalent to 2 absorbance units or 2 ppm. In this case, however, let's say you get a signal that equals only 1.5 absorbance units. So the slope of the curve is

$$\text{slope} = \frac{0.5 \text{ absorbance}}{1 \text{ ppm}}$$

That means the concentration you originally measured was not 1 ppm but actually 2 ppm. That is how standard additions works. Now, suppose that there is some background absorbance due to the sample matrix. That means the intercept of the standard additions line is not zero. If this is the case, standard additions will not compensate for background absorption. In fact, standard additions will not compensate for any other type of interference (i.e., spectral, chemical, ionization, physical) except transport-type interferences, since you cannot measure the intercept of the standard addition line. If you have background absorption and have a transport effect you have to use both standard additions and background correction.

Inductively Coupled Plasma

Nebulization

The nebulization process is probably the most important step for accurate and reproducible results.

ICP utilizes a nebulizer to transport the sample solution into the plasma. If a solution is injected directly into a plasma, it will extinguish it. If the sample is introduced as an aerosol, through a nebulizer, it can be introduced into the plasma without problem. Nebulizers can leak, clog, and generally be misadjusted. The performance of the nebulizer will vary with the viscosity of the sample, its temperature, whether there are surfactants in solution, etc. The nebulization process is probably the single most important step in ICP in getting accurate and reproducible results.

The most common nebulizer utilizes a pneumatic pump in conjunction with a concentric or cross-flow nebulizer (see Figure 2-3). Because nebulizers must have a small orifice they can become blocked (if the orifice is too large the droplet size is large and nebulization efficiency will be low). Blockage may be due to material

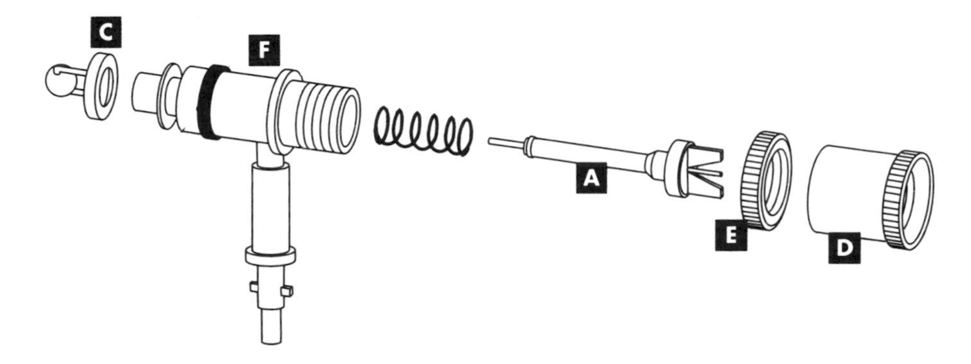

FIGURE 2-3A. Diagram of a nebulizer. Item A is the nebulizer tip. Item C is a bead which helps disperse the sample.

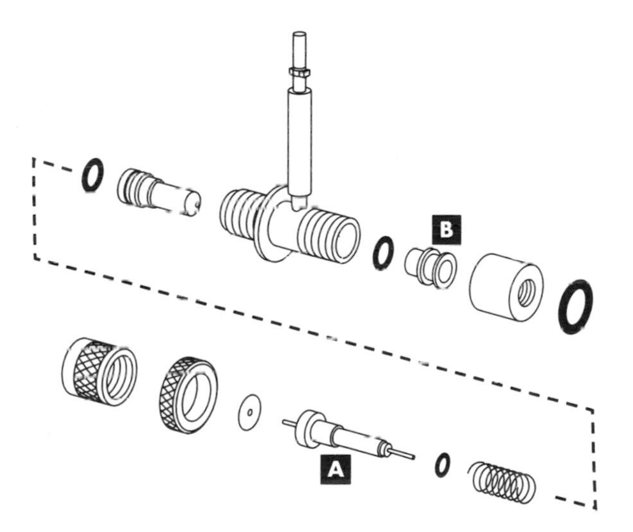

FIGURE 2-3B. Diagram of a normal nebulizer.

suspended in the sample solution or, if the sample contains high concentrations of salts, the salt can precipitate out of solution and clog the nebulizer. If the sensitivity of the system drops off and the torch seems to be acting normally, the most likely place to look for the problem is in the nebulizer. Some systems utilize a wet argon flow gas for the nebulizer which helps avoid salt precipitation. However, if the solids content of the solution is high (e.g., 1% or more) precipitation can still be a problem.

When the nebulizer is clogged the sensitivity drops off rapidly and you need to take the nebulizer apart and clean it. You can do this with a thin wire. The wire is used to remove the blockage from the capillary tube inside the nebulizer. Be careful and try not to scratch the capillary since this will change the flow through the nebulizer and may make precipitation worse. After cleaning, rinse the capillary well and reassemble. After assembly you will need to check the calibration and interelement corrections and recalibrate if necessary.

Several manufacturers offer different nebulizers for different applications. Perkin Elmer, for example, offers a high solids nebulizer that is designed for higher solids applications. Basically this nebulizer uses a somewhat larger orifice. While the nebulizer is less efficient (i.e., it uses more sample) the precipitation problem is minimized. Cross-flow pneumatic nebulizers are not very efficient; only 2 or 3% of the sample is transported to the torch because of the different droplet sizes formed during the nebulization process. Many droplets are too large to be transported to the torch and fall out of the argon gas before making it to the torch. This material ends up in the waste bottle underneath the instrument.

Note: *The waste tube will always have a loop and liquid in the loop. Liquid is eliminated from the system through a positive pressure from the instrument side of the tube and prevents air from getting back up into the instrument, which could cause plasma instability, pop, etc.*

Newer instruments may be equipped with ultrasonic nebulizers, which are more efficient than the cross-flow pneumatic type. Ultrasonic nebulizers produce a more uniform droplet size and, therefore, more of the sample is delivered to the torch (some claims are as high as 40% of the sample), which improves sample efficiency and sensitivity. Like pneumatic nebulizers, ultrasonic nebulizers are subject to clogging with high solids samples. Most people we have talked with feel that ultrasonic nebulizers are difficult to use. They are subject to frequent clogging and are not as stable as the standard cross-flow-type nebulizer. For environmental samples, most people prefer the cross-flow or concentric flow nebulizer.

The ICP Torch and Flow Rates

Most ICP torches accommodate three gas streams. These are shown in Figure 2-4. The outermost flow is typically 15 to 20 l/min of argon. This flow is the main source of argon for the plasma and is also used to cool the quartz tube that surrounds the plasma. If this flow rate is too low the torch may melt. The next gas stream is called the auxiliary flow and it typically is 200 to 300 ml/min. This flow pushes the plasma up away from the lower quartz tubes. The auxiliary flow is not necessary for the ICP-AES to work and some people don't use it at all. The last gas stream is the flow rate of gas entering the center of the torch carrying the nebulized sample. This flow rate is called the aerosol carrier flow rate and is typically 500 to 700 ml/min. The aerosol flow rate is very important since it will move the plasma up or down and thus change where the plasma is optically "sampled".

Checking for optimal aerosol carrier flow rate

Referring to Figure 2-4, the optics of the instrument will sample the plasma at different heights depending on the aerosol carrier flow rate and torch height. The

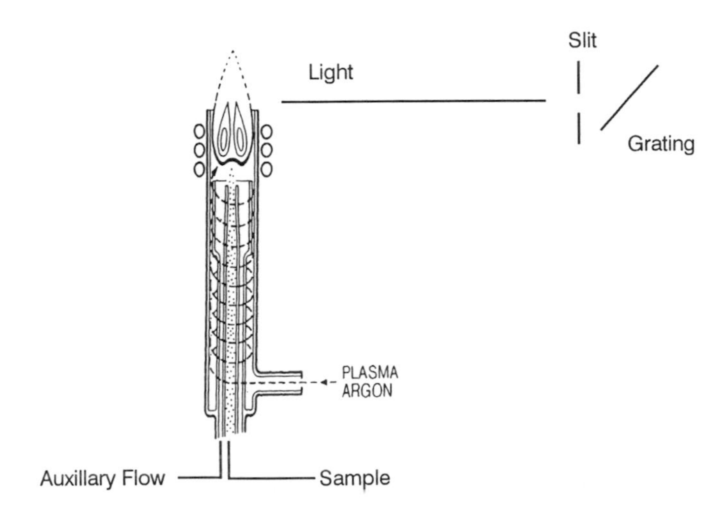

FIGURE 2-4. ICP gas streams and optical path.

higher the flow, the lower the plasma is optically sampled. By adjusting the flow rate you can determine the flow that provides the highest sensitivity for the element of interest. Normally, once the optimal flow rate is established for a given instrument it should not need to be changed unless a different torch is used or diameters of torch components change. Different laboratories use different methods to check for optimal flow rate. Here are several options you might consider:

- Set up the instrument by setting the Rf power with the minimum reflected power. This will be something like Rf power of 1100 W and reflected power less than 5 W.
- Aspirate a solution of yttrium (1 mg/ml). Adjust the aerosol carrier gas flow rate so a blue emission region extends from 5 to 20 mm (about 3/4 in.) above the top of the coil. The outer part of the plasma may appear a little pink or reddish. Record the flow rate or pressure.
- Horizontally align the plasma or optical spectrometer and aspirate a solution containing As, Pb, Se, and Tl (10 ug/ml each). Collect intensity data at 1-mm intervals from 14 to 18 mm above the top of the coil. This is called the analytical zone. Repeat the process with a blank.
- Calculate the ratio of the signal intensity for each element to the intensity of the blank for each position. Set the height for viewing the plasma to the height that gives the highest ratio for the least sensitive of the four elements.
- Prepare a solution of lead and copper (40 ug/ml). After the instrument is warmed up aspirate this solution.
- Adjust the flow rate of the nebulizing gas from 500 to 800 ml/min in increments of 25 ml/min and collect intensity data for each element. Repeat the operation for a blank and subtract the blank intensity from the element intensity. Plot these data to get the flow rate at which both elements are the highest. Calculate the signal ratio of Cu to Pb at this optimal flow rate.
- Analyze this solution every day and adjust the flow rate to give the same Cu to Pb ratio.

If flow rate changes the system will not be optimal and responses for individual elements can be expected to change (see Figure 2-5).

Sequential and Simultaneous Instruments

There are two basic types of ICP atomic emission spectrometers — sequential and simultaneous. Sequential instruments measure emission at wavelengths set by the operator in a sequential fashion. For example, lead may be measured at 220.353 nm and then chromium might be measured at 205.552 nm. The instrument moves a diffraction grating to accomplish the change in wavelength. Analysis of a sample may take up to 5 min, depending on the number of elements desired and the number of measurements (readings) taken for each element.

Simultaneous instruments are faster than sequentials, but do not permit wavelength adjustment.

Simultaneous instruments are capable of measuring 20 or more elements at one time. This requires only about 1/20th of the time of a sequential instrument, hence the basic advantage over the sequential instrument. Wavelengths are literally "set" for simultaneous instruments. **Note:** *This is true for older instruments. Newer instruments now use an eschell grating and detector array to record several spectral regions simultaneously*. The detectors (one for each element) are arrayed at precise positions. The position is dictated by the wavelength you want to observe. To change from one preset element to another element means that you have to move one or more

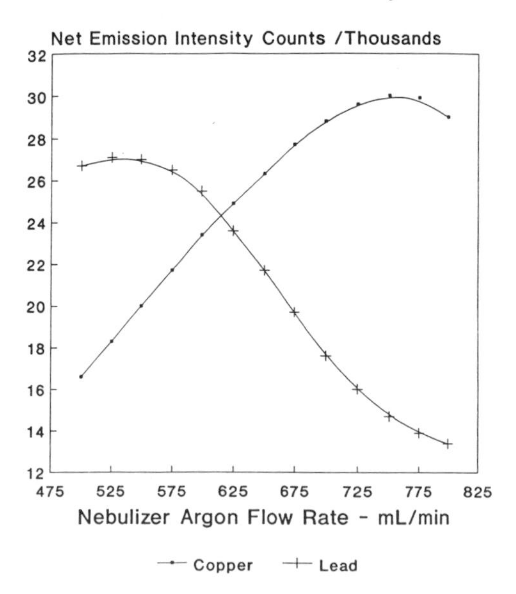

FIGURE 2-5. PB-CU ICP-AES emission profile.

of the detectors, which is time consuming. This can present problems if there is an interference at a preset wavelength. Newer instruments usually have one channel that can be set (i.e., set to any wavelength). Thus, if you suspect an interference for a fixed wavelength, you can set a different wavelength for that element without having to adjust the detector. Spectral interferences are easily compensated for with a sequential instrument simply by choosing an alternate wavelength.

Typically, when using a sequential instrument you scan over the wavelength regions of interest prior to sample analysis. This permits the operator to identify potential spectral interferences and select alternate wavelengths that are interference free. With a simultaneous instrument the operator generally assumes there is no spectral interference outside of those accounted for when using an interference check sample (see interferences and background below).

I had an interesting debate about the pros and cons of simultaneous and sequential instruments with a colleague. The colleague admitted that you were more likely to notice whether there was an interference for a given sample when using a sequential instrument. The colleague pointed out, however, that for this observation to occur, the operator would have to watch every determination, which ". . . would be about as interesting as watching bottles go by in a bottling plant." The view is that the operator would be so bored that he/she wouldn't notice the interference, or the operator would be off doing something else.

In summary, a simultaneous instrument provides data for 20 or more elements in about the same time it takes to analyze one element using a sequential instrument. When using a sequential instrument the operator is more likely to identify interferences and take appropriate action than with a simultaneous instrument.

Calibration

The ICP is calibrated by using a series of standards that covers the concentration range expected for the samples. Concentrations in the range of 0.1 to 100 mg/l are typical for ICP. Calibration is normally linear over several orders of magnitude. For example, one would expect a linear calibration for lead between 25 mg/l down to 0.25 mg/l. At higher concentrations the calibration curve may not be linear. It is common for response to fall off somewhat at higher concentrations. In these cases, it is possible to use a nonlinear calibration curve. However, most people prefer to simply dilute higher concentration samples until they are in the linear region of the calibration. This, however, takes time so a nonlinear calibration to extend the calibration range can help laboratory productivity.

You can use a non-linear calibration to cover a wider calibration range and save time

When you calibrate it is normal practice to use a concentrated standard and then to dilute that standard to cover the concentration range of interest. For example, you might start with a standard at 25 mg/l and dilute by 10 to obtain a 2.5-mg/l standard and by 10 again to get the 0.25-mg/l standard. This technique is called serial dilution and provides a linear calibration. After you establish the calibration curve it is good practice to verify the curve with an independent standard to make sure there were no errors associated with the original solution. If the original solution was really 30 mg/l then the concentrations you measure will be off by 20%. To make sure the calibration is accurate you should verify with an independent calibration solution. Typically, one would use a standard from NIST or solutions from a commercial supplier. The concentration you determine for the independent standard should agree with the certified value within at least 10%. Many methods you use will require a specific level of agreement. What do you do if all of the elements agree within the specified range but one? The first thing is to recheck the calibration and make up a fresh solution for that element if need be.

Many environmental methods call for analyzing a standard every ten determinations to ensure the results continue to be accurate. This procedure amounts to verifying the instrument calibration every ten analyses. Many instruments, especially older ones, will drift with temperature. If your laboratory is subject to temperature swings (e.g., 5°F or more) during the day, verification of the calibration is a good practice. If the value for the continuing calibration is off by more than 10% you will, at a minimum, need to recalibrate, and you should determine the reason for the change and correct it. As mentioned, temperature fluctuation can cause drift; however, instrument response may also be affected by changes in the nebulizer (i.e., clogging) or gas flow rates. ICP instruments use one or more photomultiplier tubes which do change gain over time; however, the gain of these tubes usually changes slowly. Therefore, the detector is not normally the cause of the drift.

ICP Interferences

There are four basic types of interferences possible with atomic emission spectra. There can be sample transport problems, atomic lines from one element may overlap with those of another, molecular species may form and lower the apparent concentration of the metal in the plasma or flame, and interferences may be observed by the formation of ions.

Transport problems arise from differences in sample viscosity

Of the four types of interferences, ICP is particularly subject to sample transport and spectral interferences. Sample transport can be corrected for by using an internal standard, the method of standard additions, or dilution. Transport problems occur because the viscosity of the sample is different enough from the standard that the amount of sample entering the torch is sufficiently different from the standard to cause an error. The difference in change in viscosity causes more or less sample to be nebulized relative to the standard and, therefore, a measurement error arises. There is another way to control sample transport interferences not mentioned above. You can also "matrix match" the sample and the standard. For example, if the sample contains a high concentration of salt, a standard could be made to try to match the salt content of the sample. Matrix matching, although possible, is not very practical for most environmental laboratories since it requires specific information about the matrix being analyzed. Usually, little is known about the matrix prior to analysis, therefore matrix matching isn't very efficient.

Most labs use sample dilution, the method of standard additions, or internal standards to compensate for sample transport problems. With ICP using an internal standard is much easier than using standard additions. Standard additions is time consuming since the sample must be analyzed three or more times. This is not always practical for ICP when multiple elements are being determined because each element will have to be spiked into the sample at different concentrations. Just calculating the volume of standards needed to be added to the sample will take 15 or 20 min. Therefore, use of an internal standard or diluting the sample to eliminate the transport effect are really the only practical methods to handle transport effects with ICP.

Sample Dilution

Sample dilution is an easy way to eliminate transport problems, assuming the concentration of the analyte is high enough. Here's the way it works. The sample is first determined and then diluted 5- to 10-fold and redetermined. If the two determinations agree within 10% (accounting for the dilution factor) then a matrix effect is not present. In other words, the dilution has no effect of transport. This means that the transport of the sample was about the same for the undiluted sample. Hence, no transport effect and no error due to transport. If the determinations do not agree, a transport effect is suspected. In this case the sample may be diluted further to dilute out the matrix effect (i.e., transport problem), assuming the analyte is at a high enough concentration to be determined after subsequent dilution. Basically, what is happening is the analyst is adding deionized water (or deionized water with nitric and hydrochloric acids) to the sample to get the viscosity of the sample closer to the viscosity of the standard. This means that the matrix effect is "diluted out" and the sample may be determined normally. The problem, of course, is that by the time the transport effect is diluted out the sample may be below the detection limit.

When the analyte is not detected in a sample the only thing you can do is to spike the sample (bench spike) to determine if you have a matrix effect. A bench spike is like a one point standard addition. The sample is spiked with a known amount of analyte after the initial determination. Typically, you would spike so the new concentration will be two to five times the original concentration. If the second determination is correct within 10% then no matrix (transport) effect is present. If the results do not agree the sample may be diluted and reanalyzed, an internal standard added, or the method of standard additions used.

Internal Standards

Besides dilution, using an internal standard is really the only other practical method for correcting for matrix effects when using ICP. The internal standard is an element that is added to the sample (after digestion) that is not expected to be present in the sample. Metals such as scandium and yttrium are common internal standards. The same concentration of internal standard is added to every sample and to the calibration standards. All measurements are made "relative" to the internal standard. The concentrations determined in the samples are adjusted depending on the response of the internal standard. For example, assume scandium is serving as the internal standard and is added to all of the calibration standards at 1 ppm. You construct the calibration curve by plotting the signal strength of the analyte divided by the signal strength of the internal standard versus the analyte concentration. This produces a calibration curve such as the one shown in Figure 2-6. If you have a matrix effect it will affect the transport of the internal standard as well as the analytes of interest. The internal standard is used to correct for the matrix effect. Therefore, if the signal for the internal standard goes down by 50% the signal (and ultimately the concentration) of the analyte is doubled. This approach works as long as there is no discrimination in transport of the analyte and internal standard.

Interferences and Background

As indicated above, ICP is prone to spectral and matrix (transport) interferences. Spectral interferences can take many forms. The most common involves a simple

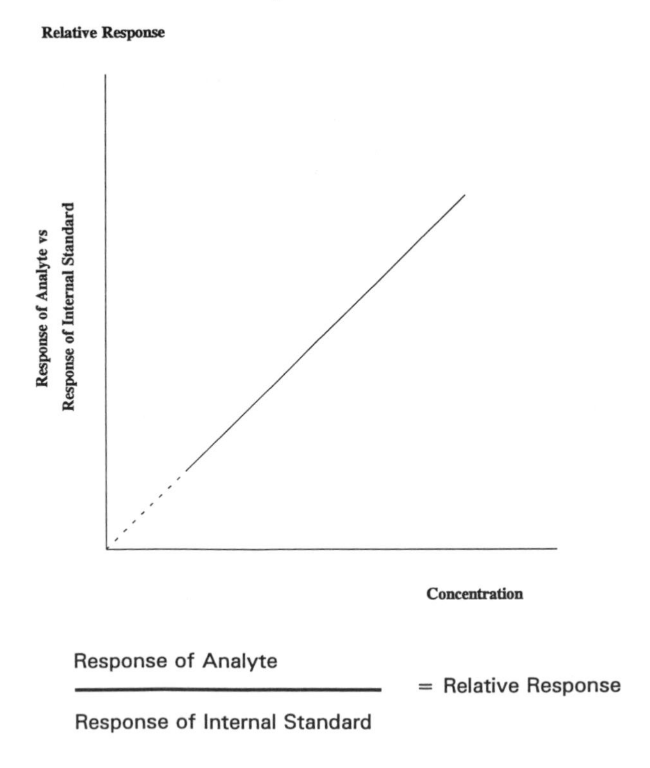

FIGURE 2-6. Calibration curve using an internal standard.

shift of the baseline on one or both sides of the spectral line. Occasionally you will observe direct spectral overlap on elements of interest.

Baseline shifts are common in ICP-AES. Baselines are often shifted upward because of the high concentrations of other elements. For example, iron will interfere with the determination of vanadium by shifting the baseline of vanadium (i.e., in the region of 292 nm) upward. Common interferences are illustrated in Figure 2-7. The line diagram in Figure 2-7A shows the vanadium line from the vanadium standard. The second line shows the effect of 1000 ppm of iron in the same wavelength region. Notice that the background is shifted higher. ICP-AES instruments measure response versus wavelength. Using this example, the signal at 292.40 nm is measured for vanadium. The signal at this wavelength is determined regardless of whether vanadium is present or not. If no vanadium is present, then the background signal is recorded as being attributable to vanadium. This signal is then compared with the signals established during the calibration and the concentration is calculated. If the background value is shifted upward as a result of iron being present, the vanadium concentration will be over-reported. In this case the 1000 ppm of iron is equivalent to about 3 ppm of vanadium.

Figure 2-7B shows how a spectral line from a sample can interfere with a determination for nickel. This type of interference would normally be missed unless

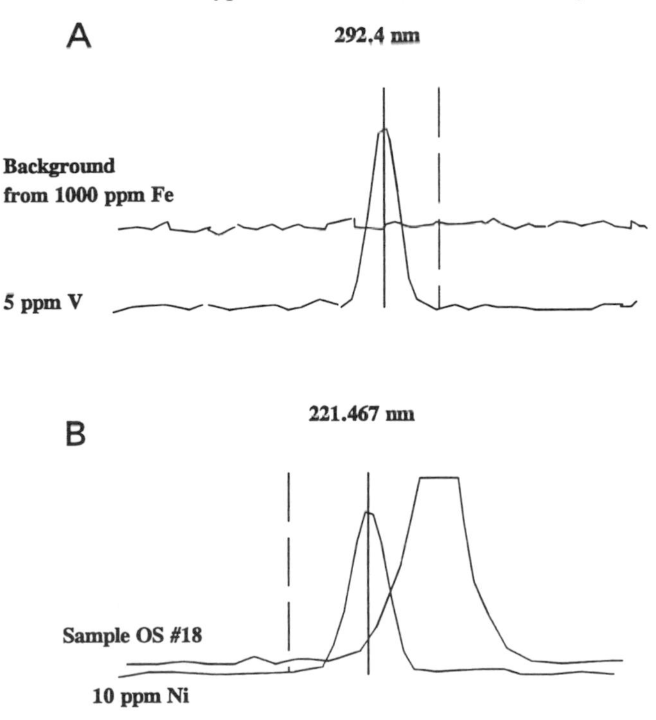

FIGURE 2-7. Interference of iron on vanadium (A) and spectral overlap on nickel (B). Sloping background in the region of Vanadium (C).

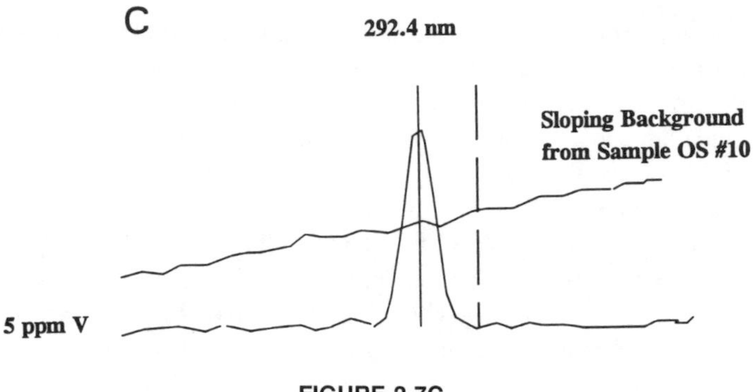

FIGURE 2-7C.

the wavelength region were scanned. This type of interference is called spectral overlap and causes overestimation of the concentration of an element in the sample. In this case we decided to change to a different wavelength to avoid the interference.

Another common type of interference comes from a sloping background, shown in Figure 2-7C. In this case, similar to the one shown in Figure 2-7A, the background signal is higher. In this case it slopes. With some instruments you can correct for this by measuring the signal on one or both sides of the analytical line and subtracting the signal. If the slope of the background is steep background correct will be difficult and you might do better with another analytical wavelength.

Interference Check Samples

One method that may be used to eliminate these types of background problems is to use an interference check sample. The interference check sample contains high concentrations (typically 1000 ppm) of elements that will likely be present in samples that will be analyzed. For example, iron, aluminum, calcium, magnesium, and sodium are all fairly common elements in environmental samples and they may interfere with other elements, especially when they are present at high concentrations. The interference check sample does not contain all of the analytes of interest. It only contains those elements that are known to interfere with other elements. The interference check sample is introduced into the ICP, measurements made for all elements of interest, and the apparent concentration for the elements is established. Using the iron-vanadium example, 1000 ppm of iron gives an offset signal at 292.40 nm equivalent to 3 ppm of vanadium. In this case we know that no vanadium is present since vanadium is not part of the interference check solution. The data systems which control the ICP can construct a linear relationship between iron and vanadium which is used to reduce the measured vanadium concentration based on the concentration of iron. If vanadium were present in a sample at 5 ppm and iron at 500 ppm then the combined signal at the vanadium wavelength would be equivalent to 6.5 ppm of vanadium (e.g., 5 ppm for the vanadium and 1.5 ppm due to the iron). The data system can correct for this since it has recorded that 1000 ppm of iron gives a 3 ppm signal at the vanadium wavelength. Using this predetermined ratio the data system subtracts 1.5 ppm from the vanadium determination resulting in a correct 5 ppm concentration for vanadium. This type of interference correction can be employed on both simultaneous and sequential instruments, and is generally required for simultaneous instruments since you cannot shift to another wavelength to avoid interelement interferences.

Background correction factors should be reevaluated

Establishing these background correction factors is a time consuming job and labs occasionally need to verify these correction factors. For example, if a torch, flow controller, nebulizer, etc., is replaced then the operator should verify these interelement correction factors. Why? Because any of these pieces of the ICP affects the transport of material into the plasma or the position of the plasma *vis-à-vis* the optical system. That means that more or less material may be entering the torch or the point of sampling the plasma has changed and, therefore, previous interelement correction factors may have changed as well.

The operator should check the system routinely to make sure the interelement correction factors are still accurate. The way to do this is to run a standard which contains the interfering elements and elements of interest and see whether the correct results are obtained. This is the normal procedure used by most labs when evaluating the instrument for use.

Flow Rate

Plasma temperature, sampling height, and flow rate are related

As discussed above, the flow rate of argon into the torch will affect the signal strength for the elements determined. This change in signal strength is due to the changes in the exact position of the plasma *vis-à-vis* the optical system. In ICP instruments you can adjust the torch position up or down and from side to side, which affects where the optical system "samples" the plasma. Also, the temperature of the plasma varies depending on its height above the torch. This means that you can, to some extent, sample the plasma at different temperatures. Some elements have stronger transitions at one temperature than at another. For example, alkali metals are ionized at relatively low temperatures. Therefore, you get the greatest sensitivity for alkali metals in the upper portions of the plasma where it is cooler. This means that there will be an optimal viewing position for each element and this position is controlled by the flow of argon into the torch. The auxiliary flow (Figure 2-4) is used to move the plasma up off the torch and can be used to control where the plasma is sampled. When you calibrate and determine interelement correction factors, you should try to maintain the same viewing position (temperature) of the plasma.

One way to do this is the operator can establish a ratio between copper and manganese or copper and lead as a test to see if the flow conditions and plasma temperature have changed in the instrument (see Figure 2-5). If the ratio of either of these signals changes, then the plasma temperature or flow into the torch has probably changed and the flow must be adjusted to coincide with the previous flow and the interference check factors validated. Using this ratio method is a quick way to insure the flow conditions (i.e., the temperature of the region of the plasma you sample) are the same from day to day and that the interelement correction factors are still accurate.

Background Subtraction

Background subtraction is accomplished by measuring the signal on one or both sides of the emission wavelength. For example, if you were determining nickel at 231.604 nm you would find that the width of the emission peak is about 0.03 nm. That means you could determine the background at 231.620 nm or at 231.589 nm, or both. You can use both sides and take the average between them if, for example, the background is sloping in one direction. The physical method of determining the background is different depending on whether you are using a simultaneous or sequential instrument.

Sequential Instruments

Sequential instruments are made to scan a series of wavelengths. Therefore, to measure the background of an emission peak, one simply scans the spectrometer about 0.15 nm on either side of the emission line (peak). In this case you are moving the diffraction grating to bring different wavelengths to bear on the detector tube. The grating can be controlled fairly precisely and the resolution with most ICP-AES instruments is about 0.002 nm. With simultaneous instruments, however, the detectors are fixed and the instrument is really not built for scanning. In this case, background is sampled off the emission peak by either moving a piece of quart glass inside the spectrometer or by shifting one of the slits.

There are a number of wavelengths that come out of the grating. These wavelengths are in a spectral region covering several nanometers. Now, when you put a piece of glass between the grating and the detector tube, as long as the glass is at 90° to the incident light, the light path is not altered. If, however, the glass is rotated slightly, the light will be bent. The different wavelengths will impact the glass at different positions. Using nickel as an example, the light at 231.604 nm will be at right angles to the glass and pass through and strike the detector tube. If the glass is rotated slightly, shorter or longer wavelengths will be shifted to focus on the detector due to the refraction through the glass. In this case, let's say the light was refracted so that 231.620 nm was focused on the detector. This is one method for determining the background signal of the analytical wavelength. The other method is to move a slit that is placed after the diffraction grating and before the detector tube. The slit can function in much the same way as the glass. By moving the slit to the right or left you allow different wavelengths to strike the detector tube. In this manner you can obtain a background reading off the analytical line. Simultaneous instruments that perform background correction by using the glass tend to be more reliable and reproducible than the ones that rely on adjusting the slit position.

Atomic Absorption

Chemical interference is more common in AA than in ARE

Atomic absorption is different than ICP-AES in that we measure the concentration of the ground state atom by its absorption of light of a characteristic wavelength to an excited state. In atomic emission we measure the amount of light given off by an excited state as it changes back to the ground state or some other excited state. Although the transition (wavelength) we measure may be the same for ICP as AA, the way we provide energy for the transition is different. As we shall see, this difference has major implications for the GFAA technique. For example, chemical interferences are virtually unknown in ICP atomic emission; however, there are significant chemical interferences with furnace and flame AA. On the other hand, spectral interferences are common with atomic emission, but are rare in atomic absorption. These differences provide the basis for using AA instead of ARE for specific elements.

Element Determination Using AA

Flame AA is best for alkali metals

Graphite furnace atomic absorption (GFAA) is used primarily for elements that have low emission and are at the low end of the spectral range (i.e., below 200 nm). Arsenic, selenium, and thallium fall into this category and are the primary elements that we determine using GFAA in environmental laboratories. GFAA is used for these metals since emission lines for these elements are weak and they are subject to interference by air at wavelengths below 200 nm. Below 200 nm the baseline emission goes up and becomes less stable. Therefore, it is easier to determine low concentrations of elements in this region by AA since it is easier to determine the difference in absorbance rather than emission. That is, the element lamp provides a strong emission source for absorption and, in the presence of air, it is easier to measure absorption of a strong emission line source rather than emission from a weak source.

Flame atomic absorption (FAA) is not used extensively in environmental laboratories due to the advent of ICP-AES. FAA does have an advantage over ICP-AES, however, when determining low concentrations of the alkali metals (especially sodium and potassium). This is due primarily to the difference in temperature between the air-acetylene flame and the argon torch. The temperature in the ICP torch is in the neighborhood of 6000 to 8000°C. At these temperatures a significant portion of the alkali metal atoms are ionized as well as in excited states. The analytical wavelength normally observed for emission is a transition from an excited state to the ground state. This means that all of the atoms that are ionized will not be detected since they emit at a different wavelength (i.e., a different transition). Thus, only a portion of the atoms emit at the desired wavelength, which in turn lowers the efficiency and raises the detection limit. The temperature of the air-acetylene flame is in the range of 1500°C. At this temperature most of the atoms of alkali metals are in their ground state and available for absorption. There is a higher efficiency for the absorption process than for emission, thus FAA has better sensitivity for the alkali elements than ICP-AES. Since these alkali elements are usually of little concern in environmental samples, determination using ICP-AES is preferred over FAA even though FAA is more sensitive. In fact, these elements are more of a concern due to potential interference they may cause in environmental samples.

Calibration

As in ICP-AES, the instrument is calibrated by analysis of a series of standards (serial diluted) and plotting the response versus the concentration of the solution. After the calibration curve is established, the curve is verified with an independent standard, just as with ICP-AES. There are, however, a few minor differences between calibrating an atomic absorption instrument and an ICP. One difference is that an interference check sample is generally not required since spectral interferences are much less of a problem in AA than in ICP-AES. Background correction is also accomplished in a different way than with ICP-AES.

Maximizing Sensitivity

Before calibration one should verify that the lamp current is set for the recommended value supplied by the manufacturer and that the lamp is correctly aligned with the furnace and monochrometer. The monochrometer is set for the specific analytical wavelength of the element and the lamp position is adjusted until a maximum intensity is read at the detector. When the intensity is maximized the lamp is aligned. In older instruments this is accomplished by turning the lamp position screws while watching the signal intensity meter for a maximum. In more modern instruments, this procedure is somewhat automated.

Many laboratories use a check sample and historical records to monitor instrument performance over time. Lack of reproducibility or short-term change in signal strength usually indicates that the graphite tube needs to be replaced. Figure 2-8 shows a diagram of a graphite tube with a L'vov platform. Graphite tubes should last between 50 and 100 firings. The actual number will depend on the sample matrix and acid mixture used. With some matrices (e.g., high concentrations of nitrate) you may only get 20 to 30 firings. Graphite tubes with a pyrolytic coating and a platform will generally last longer.

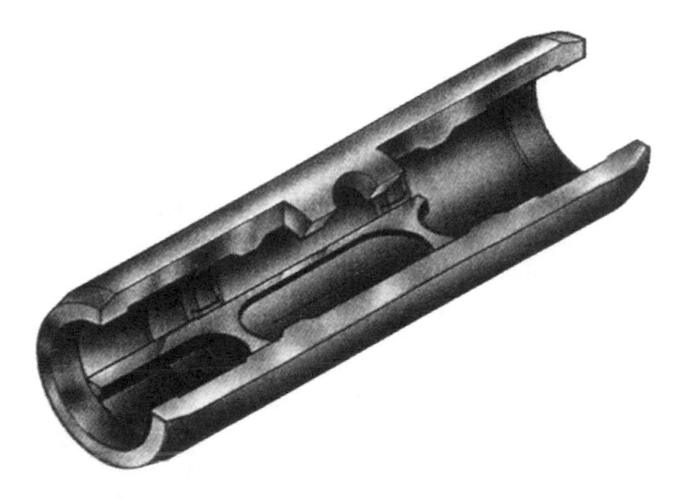

FIGURE 2-8. Diagram of graphite tube with L'vov platform.

A short-term change in sensitivity may also be caused by misalignment of the pipet arm that loads the sample into the graphite tube. If the tip becomes bent or misaligned it will cause a significant change in sensitivity. If the tip gets bumped or hits the graphite tube or hits the sample container, it can be bent or knocked out of alignment. If this is the case you can either realign the pipet tip or install a new pipet tip. Make sure the sample is being delivered at the correct position on the tube (i.e., onto the platform if you are using that type of tube).

Check samples are also useful for monitoring long-term trends. For example, suppose that 25 ul of a 50-ug/l solution of arsenic (1.25 ng) produces a certain intensity integrated signal (S). Using the same conditions over time the signal may fall to one half S. The change in signal strength might be due to dirty furnace windows, aging of the AA lamp, or the photomultiplier tube. Thus, it is good practice to keep a record of raw area counts from a standard sample to check on overall instrument performance with time.

The flow rate for the blanket gas (typically, argon or nitrogen) should be set at the recommended flow rate. This gas will keep the graphite tube from being destroyed during the analysis. For some elements (e.g., arsenic and selenium) it is recommended that this gas flow be stopped for a few seconds during the actual measurement to increase sensitivity. When the blanket gas is stopped the ground state atoms are retained in the furnace longer and provides a higher concentration in the graphite tube. This, in turn, means higher sensitivity.

Interferences

Diluting to overcome interferences is a good technique as long as the concentration is above the detection limit

There are several types of interferences possible with GFAA. There are chemical, spectral, and physical interferences. To determine if you have an interference you should analyze the sample and then dilute the sample and reanalyze. If the two results agree (within about 10% — correcting for the dilution) then an interference is not present (assuming that there is enough of the analyte present that you can still detect it after the dilution). If the two results do not agree then an interference is likely. In this case you can dilute the sample until the interference goes away. That is, as you dilute the sample you lower the concentration of the interfering component to the point where it no longer is in sufficient concentration to interfere. Unfortunately, you also dilute the analyte by the same amount and, by the time you dilute out the interference, you may also have diluted out the analyte to below your detection limit.

Sample transport effects, as we discussed with ICP-AES, are due to changes in viscosity between the standard and the sample. Since the sample is pipetted into the graphite tube, changes in viscosity are normally not a problem in GFAA. It is possible, however, to have this type of problem with certain types of samples. For instance, oil samples that have been digested using microwave or bomb techniques will be much more viscous than the acid digest from water or solid samples. Samples that contain high solids may also present a viscosity problem. In this case you may need to dilute the sample sufficiently to eliminate the viscosity problem.

Chemical interferences come in two varieties. There may be interference due to the formation of a stable molecule with the atom of interest or there may be absorbance from the matrix. In the first case, the metal is tied up in a molecule. Since the metal in the molecule will not absorb the light at the analytical wavelength you get a low result. In the second case the matrix absorbs light at the analytical wavelength and you get a high result. Matrix modification will generally help in eliminating the first and, to some extent, the second problem. Background correction is also very helpful in correcting for the second problem.

Before discussing the use of matrix modifiers, we should mention the steps used during a typical graphite furnace analysis. Typically, four to six steps make up the temperature program used during a furnace analysis. During these steps the temperature of the graphite tube is quickly changed and the flow rate of gas in the tube may also be changed. After the sample is deposited in the tube, either on the tube itself or on a platform in the tube, the temperature is raised to 100 to 120°C to dry the solution. Sample sizes up to 100 ml may be used in the tube and up to 50 ul on the platform. Drying usually takes about 1 min. Next the temperature is ramped up to about 700 to 800°C while gas is flowing through the tube. This step is sometimes called pyrolysis or charring or ashing. During this time you try to remove as much of the matrix as possible while leaving the analyte of interest behind. After the char step the temperature of the tube is brought up quickly to atomize the sample remaining in the tube. The temperature will depend on the analyte and its form in the tube. You should use as low a temperature as possible and still atomize the sample. During this time the gas flow through the tube may be reduced or shut off completely. After atomization the temperature may be raised even further to clean out the tube before the next sample is deposited. The temperature cycle of the graphite furnace is shown in Figure 2-9.

Matrix Modification

When using matrix modification the ideal situation is to add one or more compounds to the graphite tube that will remove everything in the sample matrix except the target metal during the charring portion of the GFAA analysis. That is, you can either add compounds that will cause the sample matrix to become volatile enough to be removed during the char phase or you can add compounds that decrease the volatility of the target metal. In either case you increase the effectiveness of the char to remove the sample matrix from the target analyte. The classic example is the removal of NaCl by adding NH_4NO_3. NaCl decomposes at 1400°C so it will not be removed during the char phase. However, NH_4Cl decomposes at 330°C and $NaNO_3$

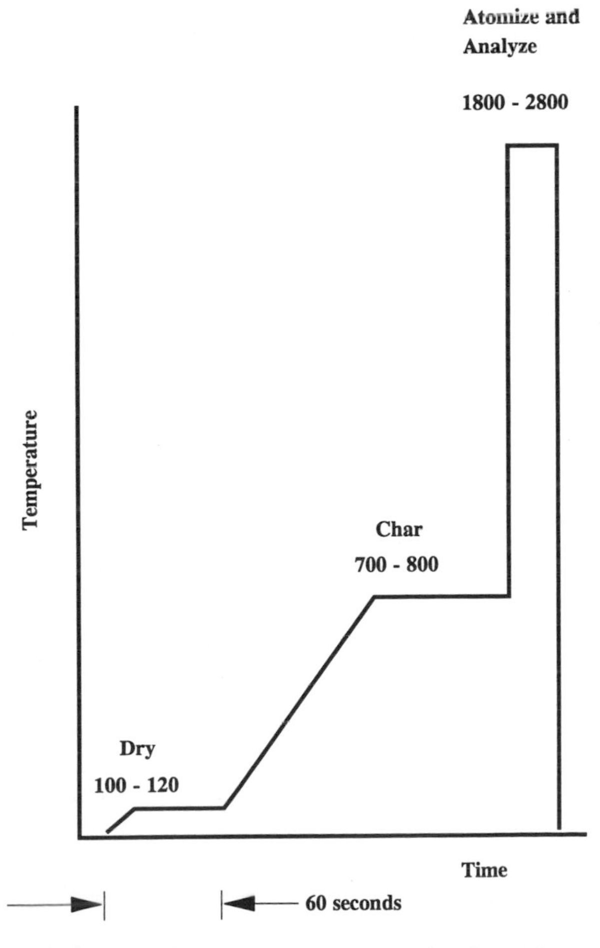

FIGURE 2-9. Furnace temperature as a function of time.

decomposes at 380°C. When you add excess NH_4NO_3 in the presence of NaCl, you form NH_4Cl and $NaNO_3$.

When analyzing for selenium most methods call for the addition of $Ni(NO_3)_2$. The Ni causes the volatility of the selenium to increase. Without the addition of nickel most of the selenium would be lost during the char cycle since selenium is a fairly volatile metal. There are many other examples of the use of matrix modifiers in GFAA. The specific methods you use will have directions for modifier use.

Zeeman Background Correction

Although matrix modification is a good technique it will often not be able to remove everything in the sample matrix. Therefore, background absorption can still present problems, even when matrix modifiers are used. The matrix components that are not removed during the char cycle are volatilized along with the target metal and may absorb light over a wide range of wavelengths. If these components absorb light at the analytical wavelength, you will have a difficult time distinguishing the analytical signal above the background. You can get around this problem by background correction. Zeeman background correction is one of the most common background correction techniques in use today for GFAA.

Without getting too involved in the theoretical aspects of Zeeman correction the technique works as follows. Referring to Figure 2-10, the ground state of the target atom absorbs light at the analytical wavelength and an electron is promoted to an excited state. If other matrix components also absorb at that wavelength you get a high reading; that is, you overestimate the concentration of the target analyte. In the presence of a strong magnetic field, however, the energy level of the excited state splits into two energy levels, one higher in energy and one lower. Now, an electron from the ground state atom can no longer be promoted to the original excited state

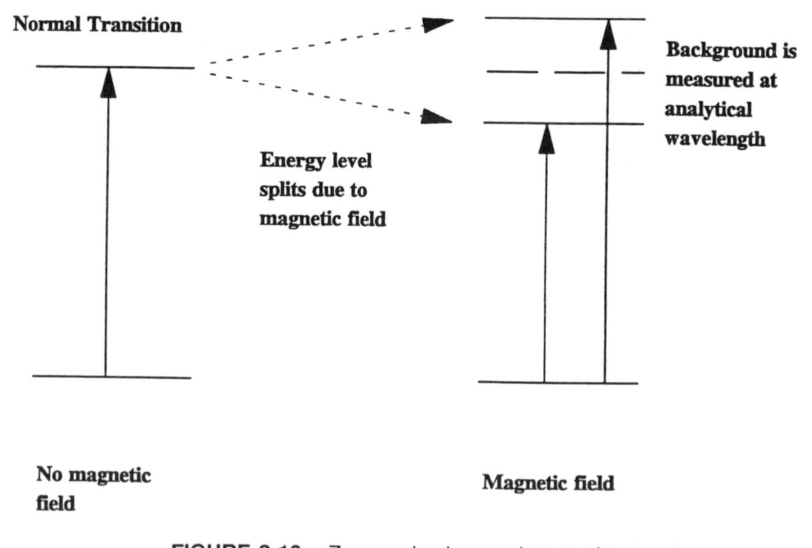

FIGURE 2-10. Zeeman background correction.

since that excited state no longer exists. The original excited state has become two other states that have different energies. Since the two new excited states have different energies than the original excited state, they absorb light at frequencies that are different from the analytical wavelength. Remember that the analytical wavelength is the exact energy required to promote an electron from the ground state to the original excited state. Therefore, no absorption by the ground state atoms can occur because the wavelength provided by the AA lamp is incorrect for the two new energy levels. When the magnetic field is on only the absorption due to the background is measured and the measurement occurs at *exactly* the analytical wavelength. This is why Zeeman background correction is such a powerful technique.

CHAPTER 3

The Organic Laboratory

Most organic laboratories, especially environmental, are divided into three or four sections (Figure 1-1). These sections represent the basic operations and organization of the lab. Generally, there is a sample preparation section, which prepares the samples, including quality control (QC) samples for analysis; a gas chromatography section, that determines pesticides and PCBs; and a gas chromatography-mass spectrometry section which performs GC/MS on prepared samples.

SAMPLE PREPARATION

Every sample, with the exception of aqueous samples for volatile analysis, requires some form of sample preparation. The way the sample is prepared will depend on the determination method, which is directly related to the analytes, required detection limits, and the matrix.

Matrices — An Introduction

Aqueous and Leachate Samples

In the normal course of analyzing samples you can expect to see just about anything. Water samples from various sources (wastewater, surface water, drinking water, groundwater, etc.) are relatively common and generally represent the best-characterized sample matrix. Included in this category, the lab may be required to analyze aqueous leachates which are generated as part of sample analysis. The leachate is an aqueous extract of a solid, semisolid, or possibly an organic material that contains acetic acid, an acetic acid buffer, and/or dilute mineral acids (e.g., nitric and sulfuric acids). These samples are extracted with solvent (usually methylene chloride) and concentrated.

Solids, Sludge, and Soil

Solid samples, sludge, and soils are another common category. Solids and sludges are typically wastes from an industrial process. These types of samples are also extracted with a particular solvent system, depending on what the composition of the sample material is and what compounds one is interested in determining. Normally we use either methylene chloride or a mixture of acetone and hexane for these extractions.

Oils and Organics

Oil is another category of sample that we find with some regularity. Generally one can simply dissolve the oil in methylene chloride and analyze the oil directly. Because the oil is composed mostly of hydrocarbons that will interfere with normal gas chromatography, the detection limits for many compounds will be higher than with aqueous samples. PCBs are a favorite target analyte in oils. In this case the oil may be dissolved in a non-halogenated solvent and analyzed directly, or if need be (and it almost always is needed) the sample is cleaned up by washing with sulfuric acid and separated from interferences on a florisil or alumina column.

Air

Air samples are normally taken in one of two ways. One way is to use a canister that has been evacuated, taken into the field, and opened to collect the sample. The canister is then shipped to the lab for analysis. Once at the lab the contents of the canister are swept into a purge unit or possibly onto the head of a chromatographic column. The other way to take air samples (and stack gas samples as well) is to draw the air, at a given sampling rate, through a series of traps and condensers. The traps are shipped to the lab for extraction and analysis. A variation on this is to use cryogenic traps. The problem with both of these approaches is the compounds that you actually determine in the lab may not be the ones that were trapped in the field. Chemicals can cross react with one another while trapped on a resin or may react with the stainless steel of the canister.

Sampling Heterogeneous Materials

Samples may be homogeneous or heterogeneous. Homogeneous materials are relatively easy to deal with. Heterogeneous materials may require additional processing and/or multiple analyses.

There are only three choices for the laboratory when dealing with heterogenous materials — you can:

- Separate the material into homogeneous phases or components and analyze each
- Homogenize the sample to get a homogeneous mixture — from which a subsample is removed and analyzed

- Analyze the entire sample

The choice of which of these options to employ is up to the laboratory and will be influenced to a large degree by the analytes of interest. Either or all of the three options may be used for samples that will be analyzed for metals, anions, or semivolatile organic compounds. Because volatile components will escape during processing, however, separating phases may not be viable. Each of these options is discussed in the following sections.

Volatile Samples

There is very little sample preparation required for aqueous samples when performing a volatile determination. Solid samples and oils, on the other hand, continue to present a challenge for volatile analysis. There are several techniques that can be used to analyze volatile constituents that include direct injection, purge and trap, and head space analysis. The most basic, and the one method often overlooked, is direct injection. This technique works well for samples expected to have concentrations of components in the range of 10 to 100 ppm or higher.

Direct Injection

Samples that have an odor, phase separation, or other evidence of gross contamination should be screened (via GC-flame ionization detector [FID]) to determine the appropriate dilution factor for subsequent analysis. If you don't screen your samples you can contaminate your analytical instruments and lose a significant amount of analysis time due to instrument cleaning. For aqueous samples, 1 µl of the sample can be injected directly onto a short bonded phase capillary or packed column. If the sample is aqueous you can fill a 5-ml gas tight syringe with the sample at this time, assuming you will analyze the sample by purge and trap later. Oils can be diluted with hexane or other solvent and 1 µl injected. Solids must be extracted with either methanol, tetraglyme, or polyethyleneglycol. The methanol extract can be analyzed directly. You can also use the methanol, tetraglyme, or polyethyleneglycol extract in a sparge unit for subsequent purge and trap analysis (see the next section).

Screening Samples

The direct inject technique is very simple and fast and saves several steps over the more common purge and trap. If you use an FID to screen samples you will need to inject 1 to 10 ng to get a significant response. (The response can be adjusted by using the attenuator and amplifier. We find that an FID is quite stable when operated with amplification and attenuation such that 10 ng will give a full-scale response.) If you inject 10 ng in 1 µl of the sample or extract the component is 10 ng in 1 mg (i.e., 1 µl) or 10 ppm.

Purge and Trap

Purge and trap is a little more involved than direct injection but offers lower detection limits, usually in the 1 to 10 ppb range for aqueous samples. The lower detection limit is achieved because the sample is concentrated on the trap. Thus, all of the sample (or almost all of it) enters the gas chromatographic column for analysis. If you assume that a given detector, for example an FID or mass spectrometer, needs about 10 ng of a compound for accurate identification and quantification, then the sample must contain 10 ng/5 ml of sample (assuming a 5-ml sample size), or 0.2 µg/l which is 0.2 ppb.

Normally the technique proceeds as follows:

- A sample is poured in the barrel of a 5-ml glass syringe that is fitted with a gas tight valve.
- When the syringe is filled the plunger is inserted, the valve opened, and the volume of the sample is adjusted to 5 ml.
- If you only have one vial of the sample (typically 40 ml), it's a good idea to fill a second syringe and close it. This syringe is used in case something goes wrong with the first analysis, or if a dilution is needed.
- If you are using internal standards or surrogate compounds for your analysis, take up 1 to 10 µl of internal standard solution and/or surrogate solution into a 10-µl syringe using the solvent wash technique. **Note:** *When using the solvent wash technique you rinse the 10-µl syringe several times with solvent and expel all solvent from the syringe. Remove the needle from the solvent wash and draw up the plunger far enough so you can see the solvent that was left in the needle.* Typically about 1 µl of solvent will be in the syringe at this point. Push the plunger down so the bottom of the solvent in the syringe is at the 1 µl-mark. Insert the needle into the sample solution and draw up the desired amount. For example if you want 1 µl of solution pull the plunger back so the solvent front is at the 2-µl mark. Remove the needle from the solution and pull the plunger back to inspect how much sample liquid is in the syringe. In this case you will have an air space, 1 µl of sample, air space, 1 µl (approximate) of solvent, and then the plunger. In effect you bracket the sample with air and thus you are able to determine exactly how much sample is being injected (see Figure 3-1).
- Inject the contents of the 10-µl syringe into the 5-ml syringe by opening the gas tight valve and injecting through the valve directly into the sample.

What to do about Acetate Buffers

- Screw the 5-ml syringe onto the purge unit and deposit the entire sample into the purge unit and begin purging the sample.

If you are analyzing a leachate sample that contains an acetate buffer, you will introduce significant amounts of acetic acid onto the trap and subsequently, into your GC column. Excess acetic acid will significantly reduce the lifetime of the column and reduce the separation efficiency of the column. One way around this is to make sure that all of the acetic acid is in the acetate form prior to purging the sample. To

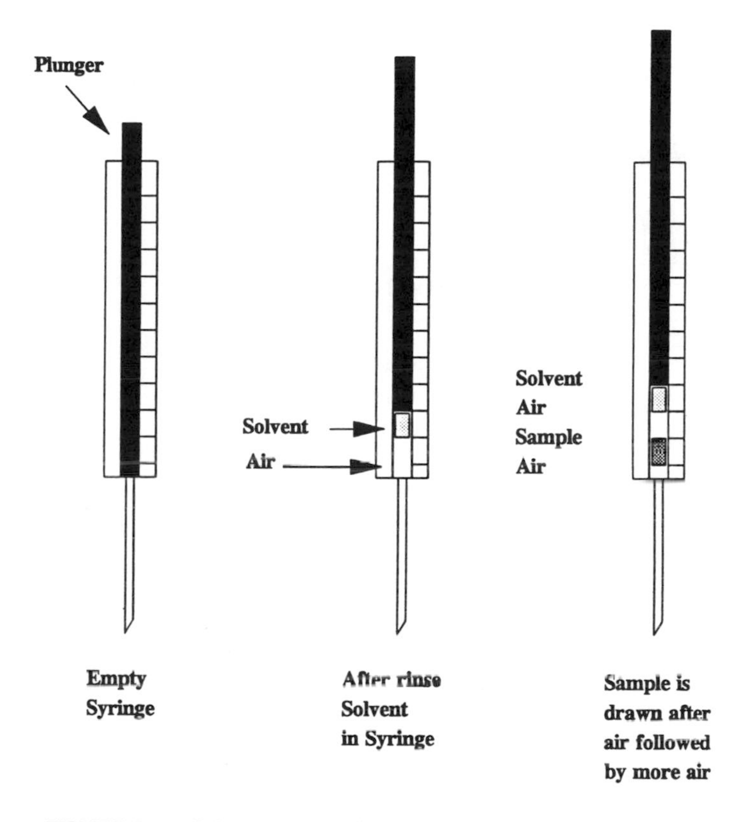

FIGURE 3-1. Solvent wash technique for accurate injection volume.

do this adjust the pH of the portion of the leachate that will be purged with sodium hydroxide solution. If you use a 5-ml sample of leachate it will take approximately 0.2 ml of 1 *N* NaOH solution to adjust the pH to between 7 and 8. The increased pH will have no effect on the purging efficiency and will not cause reactions of the target analytes. In fact, some studies indicate that, because of the increased salt concentration, purging efficiency is improved slightly.

Samples that foam

Some samples foam quite a bit due to surfactants or other contaminants present in the sample; however, this is generally of little consequence. If foaming is extensive, the best option is to dilute the sample with organic free water by taking a smaller sample and making up the volume to 5 ml. To do this follow the same procedure as above, but use a smaller sample, say 1 ml followed with 4 ml of organic-free water that has been spiked with standards and surrogates. This normally will solve the foaming problem; however, the detection limit will be higher by the dilution factor.

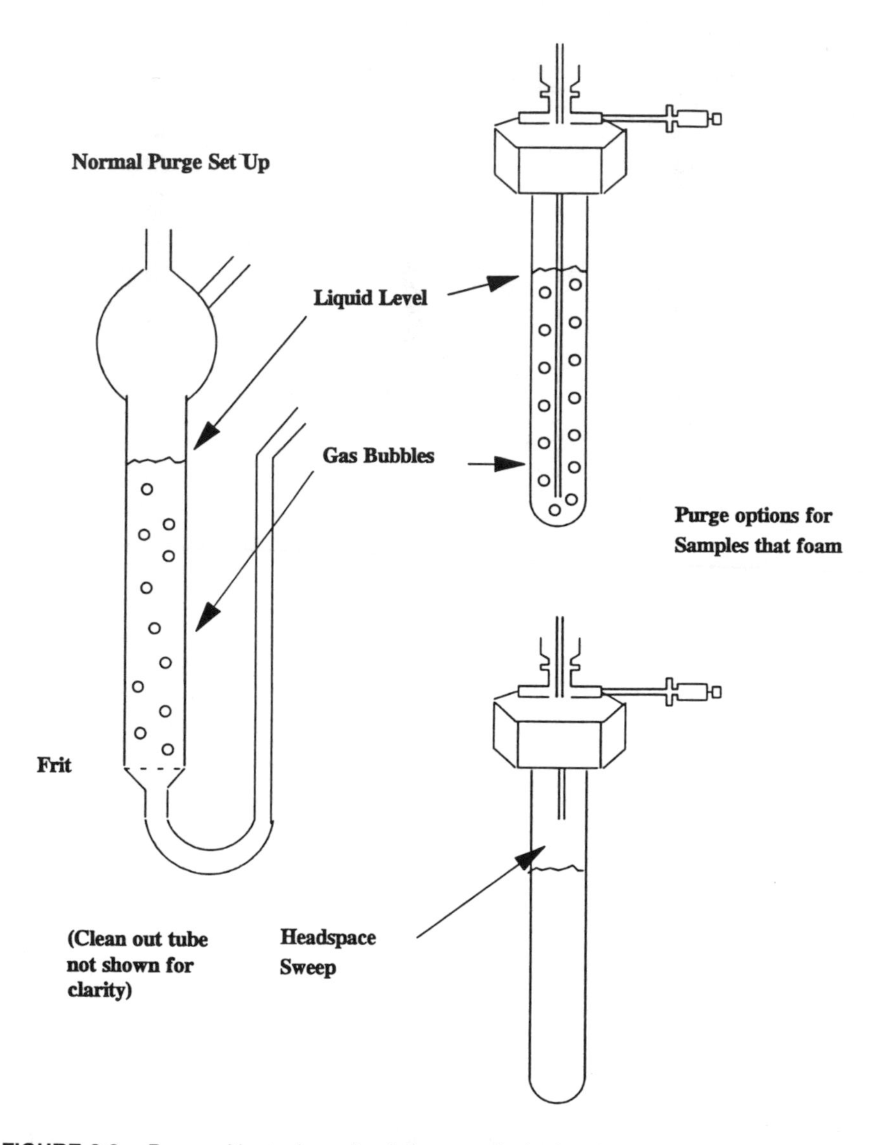

FIGURE 3-2. Purge with steel needle. Adjust needle for headspace when samples foam.

Another option is to purge the sample with a straight tube instead of through the frit used in normal purge tubes (see Figure 3-2). In this case the sample is placed in a tube (similar to a test tube) and the purge gas is introduced through a steel tube which is placed below the surface of the sample. The purge gas bubbles through the sample. However, since the bubbles are larger than the bubbles produced through the frit of the standard purge unit, foaming is greatly reduced.

Volatile components in Solid Samples

If foaming is still a problem then you can use a headspace sweep. In this case the same set-up is used as described above, except the tube is held above the surface of the liquid sample. This type of purge is not as efficient as the standard purging devices; however, you can still get about 70% or more of the analytes out of the sample using this technique. This technique is especially useful when you have samples that foam excessively or with solid samples.

Solid samples, and aqueous samples that contain solids, present an interesting analytical problem. Aqueous samples that contain small amounts of suspended solids (e.g., a few percent) can be analyzed directly in the normal purge unit as described above. A small amount of solids in the purge unit does not affect results and the solids can normally be removed through the clean out line just like any other aqueous sample. Sometimes a solid can be added directly to the purge unit if it is fine enough. For example, fine soils or a moist sludge can be added directly to the purge unit or placed into a purge tube. Transfer should be done as quickly as possible to avoid possible losses of volatiles. After the transfer normally 5 ml of water is added that includes the internal standards and surrogate compounds and the purge and trap proceeds as usual. Solid samples, including soils, can be extracted with methanol or some other solvent (e.g., tetraglyme or polyethyleneglycol that has been purified by heating it while applying a vacuum) and a portion of the extract is added to the purge unit. The only problem with this approach is that the amount of "sample" that you can purge is limited. Practically, you can only introduce between 5 and 25 µl of a methanol extract into the purge chamber before you overload the early part of the chromatogram. This presents problems for GC-FID, GC-electron capture detectors (ECD), and GC-MS. GC with conventional detectors will respond to the massive amount of methanol (i.e., parts per thousand) and you will not be able to detect the very early eluters (dichlorodifluoromethane, chloromethane, vinylchloride, bromomethane, etc.). The methanol will cause a solvent tail that makes compound identification in the early part of the chromatogram difficult. The effect will be more severe with larger amounts of methanol (sample) added to the purge unit. One way around this problem is to use the higher boiling extraction solvents (e.g., tetraglyme or polyethyleneglycol), however, their purity should be evaluated prior to use and purified if necessary.

Headspace is a good sample screening method

Make sure you know what sample basis the results will be reported in (i.e., on a wet or dry weight basis). We tend to favor reporting results on an "as received" basis, which is generally considered wet weight.

Headspace

Another alternative that may be used for volatiles is to remove some headspace from the sample container by using a headspace sampling instrument or a gas tight syringe and injecting the headspace into the gas chromatograph. Sample temperature and matrix composition affect the concentration of volatile analytes in the headspace. Therefore, headspace analysis is semiquantitative unless you construct a calibration curve for that specific matrix and use an internal standard. Headspace is a useful technique for determining whether volatile constituents are present in a sample above a threshold concentration.

Currently there are several headspace sampling instruments on the market that are designed to be part of an integrated GC system. These systems eliminate the temperature variability and can be used in either a static or dynamic headspace mode. In the static mode a portion of the headspace (constant temperature) is removed and injected into the gas chromatograph. In the dynamic mode, the headspace is "swept" for a period of time and deposited on a trap. The trap is then heated and the sample is deposited on the GC column. This headspace sweep technique is very similar to the purge and trap headspace sweep technique described above.

Be careful of Cross Contamination when using Ultrasonic Extraction

Semivolatiles

Ultrasonic Extraction

Ultrasonic extraction is an effective means of extracting solid samples. The sample is dispersed into the extraction solvent (usually methylene chloride or a mixture of methylene chloride, acetone, or hexane) by using an µltrasonic disrupter. Typically the Ultrasonics Inc. Model W-385 Sonicator with 475 W of pulsing capacity with 3/4 in. tapped high gain Q-type disrupter horn is used with a 50% duty cycle. The ultrasonic disrupter horn can cause cross-contamination between samples, especially if one sample is highly concentrated and the other is a low concentration sample or a blank. The horn gets contaminated and contaminates the next sample if it is not adequately cleaned. The horn should be rinsed thoroughly between each sample by placing in a small amount of solvent and turning on for a minute or two. This will effectively remove any residue left on the tip of the ultrasonic extractor.

Ultrasonic extraction is fairly rapid (1 to 3 min for each extraction) and is a good technique. However, it requires a technician to extract each sample, filter or decant the solvent and reduce the extract volume. Most labs add Na_2SO_4 during the extraction to avoid the extra drying step. The sodium sulfate absorbs water during the extraction and cuts the sample preparation time by eliminating the drying column. The sodium sulfate also aids in dispersing the sample during extraction. For unattended extractions (perhaps during an evening shift) soxhelet or continuous liquid-liquid extraction may be more convenient.

Sodium Sulfate absorbs PAHs

Soxhelet Extraction

Soxhelet extraction is the standard method used by most laboratories for solid samples. Most methods call for placing the sample into the extraction thimble and then adding some Na_2SO_4 to the thimble or mixing the Na_2SO_4 with the sample in the thimble. The Na_2SO_4 is used to help dry the extract as it percolates through the sample.

Recent studies have shown that Na_2SO_4 absorbs polynuclear aromatic hydrocarbons (PAHs) and reduces the recovery of these compounds significantly. Therefore, it is better to minimize the use of Na_2SO_4 or leave it out entirely if you want good recovery for PAHs.

Liquid-Liquid Extraction

Liquid-liquid extractions, employing either a separatory funnel or continuous liquid-liquid extractor, are routinely used. The advantage of the continuous extractor is that formation of an emulsion is less likely than with the separatory funnel.

Many laboratories add NaCl or Na_2SO_4 to the sample prior to extraction to increase the extraction efficiency for nonpolar compounds. The dissolved salt helps to increase the polarity of the aqueous phase and helps to "salt-out" the organic compounds. Water that is made about 2% by weight in salt will increase the recovery of organic compounds by about 10% for compounds in the 20 to 200 µg/l concentration range.

PCBs and Pesticides

PCBs are relatively easy to determine in a variety of matrices. A water matrix is straight forward and the detection limit is quite low. In order to identify and quantify a PCB you need to inject about 10 ng of the PCB on the GC column, assuming an electron capture or electrolytic conductivity detector. If the final extract is 1 ml and you inject 1 µl, then there would have to be 10 µg extracted from the sample. If the sample is 1 l of water the detection limit is, therefore, 10 µg/l or about 10 ppb. If the sample is an oil the detection limit is about 5 ppm due to necessary sample dilution.

Salt Increases recovery in Liquid-Liquid Extractions

Water, soils, and solid samples are extracted into methylene chloride the same way as you handle semivolatile extractions. The extracts are dried using sodium sulfate and may be further cleaned up using gel permeation or cartridge clean-up techniques. The final extract must be solvent exchanged into hexane prior to analysis using ECD or Hall detectors since the methylene chloride will overload either of these detectors. From this a 1-µl injection is made in the splitless mode.

Wash oil samples thoroughly

Oil samples are more difficult. The way we handle oils is to dissolve the sample in hexane — typically, 1 g of sample to 10 ml of hexane. The hexane solution is then washed with concentrated sulfuric acid. Usually 2 ml of sulfuric acid shaken with 10 ml of the hexane solution will remove most of the interfering components. The procedure we have used is to simply put 10 ml of the hexane extract into a 40-ml VOA vial and then add 2 to 4 ml of concentrated sulfuric acid. Shake the vial and the hexane solution will lose most of its color (the impurities go into the sulfuric acid). The hexane is decanted or removed via Pasteur pipet. At this point the hexane extract will usually be clean enough to analyze directly by gas chromatography. However, we have found that a second clean-up stage is usually advisable. This is accomplished by adding about 1 ml of the hexane extract on to a florisil cartridge and eluting with 2 ml of hexane. The PCBs are quantitatively recovered in the eluate. The eluate is concentrated to the original 1 ml and 1 µl of the extract is injected. The results of washing oil samples with sulfuric acid and florisil clean-up are shown in Figure 3-3.

SAMPLE DETERMINATION

Gas Chromatography

Most of the methods in use today require gas chromatographic separation prior to the actual determination. The following is a brief description of gas chromatographic set-up and operation.

Injectors

The injector is probably the most important part of the gas chromatograph and probably accounts for about 80% of the problems you will encounter in gas chromatography. Most of the work performed in an environmental laboratory is done in the splitless mode. In this case the injector is lined with a smooth glass insert. Several types of injector inserts are shown in Figure 3-4. The capillary column is threaded into the injector and placed at a height in the injector so that the syringe needle, when inserted through the septum, will be 1 to 2 cm above the end of the column. Make sure the end of the column inserted into the injector is cut at a right angle to the column. If the end of the column is jagged or at a different angle it will affect the flow of material into the column and give less precise results.

Depending on the column, a 1- to 2-µl sample is injected. Just before the injection takes place the split flow from the injector is turned off and the injection is made. See Figure 3-5 for a diagram of a typical injector. After the injection the solvent and solute vaporize and are deposited onto the column, which is held at a temperature just below the boiling point of the solvent. The solvent condenses on the column and traps the solute that enters the column. After a predefined interval, the split valve opens and the remaining solvent and solute are swept out of the injector. Studies have shown that most of the solute is deposited on the column and little is lost when the split valve opens, as long as the split valve remains closed long enough. How long is long enough? The longer the time before the split valve opens the better — more material is deposited on the column. There is a practical limit to this time, however, since the longer the time the longer the solvent front during which the chromatogram is obscured by the solvent. Thus, the time between the injection and the opening of the split valve is a trade-off. Normally, the time is between 30 and 60 sec. One of the most common problems in the lab is the split valve opening too soon or being set in the wrong position at the beginning of the analysis. It is easy to confuse the valve position (i.e., thinking it is closed when in fact it is open). Most gas chromatographs permit time programming the split valve during the chromatographic run. This often adds to the confusion. The cycle we use is as follows:

The split valve timing cycle

- The split valve is closed from the preceding chromatographic run. So when the injection is made the valve is closed.
- The injection is made and the valve is opened after 1 min. During the injection the column oven is maintained between 40 and 50°C depending on the solvent being used. **Note:** *The room temperature will affect the lowest temperature at which you will be able to hold the GC oven.* If the temperature in the room is 30°C then it will be difficult to hold the oven at a temperature of 40°C or less. Some laboratories use liquid nitrogen to cool down their GC ovens between injections. If liquid nitrogen is being used in your lab you can hold the GC oven at even lower temperatures. The rule of thumb is to cool the oven to 10 to 20° below the boiling point of the solvent being used for the injection.
- The sweep valve is closed again after 1 minute and remains closed until the next injection.
- Figure 3-6 shows the split valve timing cycle.

Make sure the split valve is closed when the GC is not in use.

This procedure works well and limits the amount of gas that is lost through the split valve. If the split flow is set to 50 ml/min and you leave it on during the run only 2% of the gas you buy is used to perform chromatography. You should always make sure the split valve is closed when you are not using the GC. If you keep the GC oven temperature low and the split valve closed for an extended period (e.g., overnight) you will have to program the GC through at least one temperature program to remove interfering material that gets deposited on the column. These impurities may come from the carrier gas, septum, or gas lines. If the sweep valve is left on there will not be much of a build-up of material; however, you will lose a substantial amount of gas through the sweep line. In my opinion, it is better to leave the sweep closed and either program the oven at the beginning of the day or keep the oven temperature high enough so material does not build up on the head of the column rather than lose all of your carrier gas out the sweep valve. Considering that a tank of helium can cost

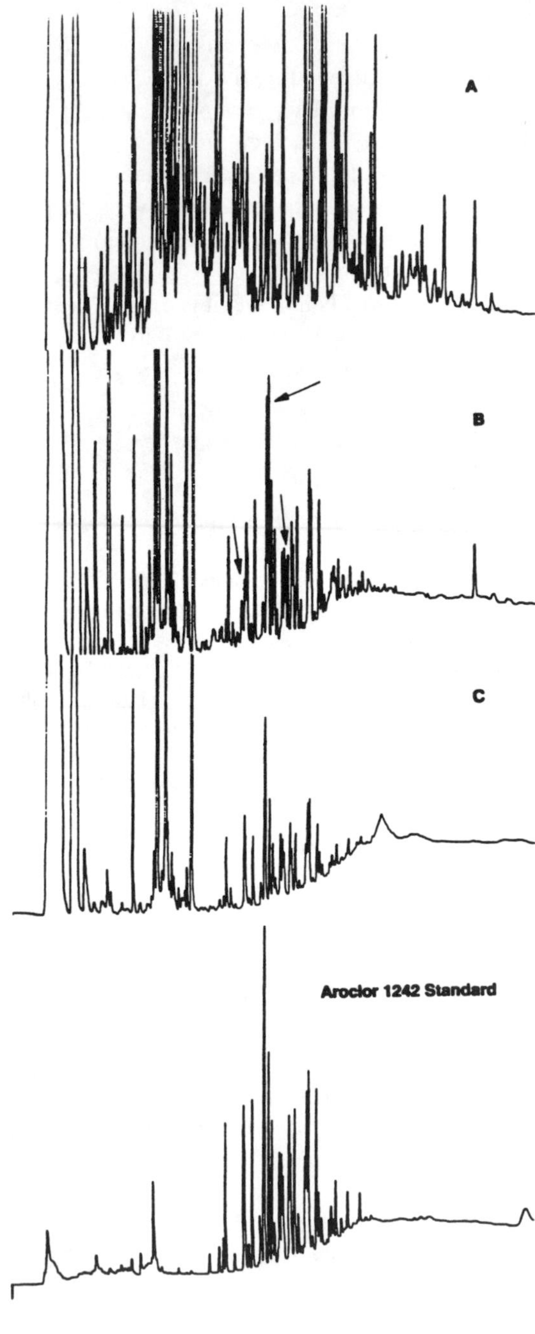

FIGURE 3-3A.

FIGURE 3-3. (A) Chromatograms from a sample of transformer oil 1. A is from the untreated oil; B is the oil after sulfuric acid treatment; and C is the oil after both sulfuric acid and florisil treatment. The arrows in Figure 3-1B show the interferences removed by the florisil. (B, on page 80) Chromatograms from a sample of transformer oil 2. A is from the untreated oil; B is the oil after sulfuric acid treatment; and C is the oil after both sulfuric acid and florisil treatment.

between $190 and $320 (depending on the grade of helium you buy) and you can easily go through one tank a week if you keep the split valve open most of the time, you can save a substantial amount over the course of a year by keeping the split valve closed.

Solvent Tailing

If you get excessive solvent tailing then the split valve is set wrong or there is a blockage in the sweep lines. The solvent front should come down rapidly within 2 to 3 min after the injection, assuming a 30-m column. If you get tailing check the split valve and the septum sweep valves.

Injector Liners

Another good technique to use is to deactivate the injector liner. It is good practice to start each day, or even each shift, with a freshly cleaned and deactivated injector liner. Changing the injector liner ensures that responses for analytes remain constant over longer periods of time, the need for frequent recalibration is eliminated, and subsequently you can analyze more samples (for profit) in a shorter period. You should have about six liners on hand for each GC and you should clean and deactivate them in batches. The deactivation procedure is as follows:

Deactivate Injector Liners

- Clean the injector liner using detergent and hot water. Scrub out residue using a small circular brush. You can get the brushes from your chromatographic supplier. Rinse the injector with water and dry in an oven at 105°C overnight. You might also consider buying lengths of glass with the same inner and outer diameters as the injector inserts and cut them to length. If this is the option you use you can just throw away the old liner and eliminate cleaning. **Note:** *Some laboratories (and GC manufacturers) prefer using quartz inserts, claiming quartz is more inert.*
- Prepare a derivatizing solution by combining a 3:1 volume:volume mixture of hexamethyldisilazane (HMDS) and trimethychlorosilane (TMCS).
- Cool the inserts in a desiccator to room temperature and then place the inserts in the HMDS/TMCS solution.
- Store the inserts in the derivatizing solution until you need to prepare a fresh batch of inserts.
- Remove the inserts from the solution and shake off the excess. Place the inserts into a drying oven and bake at 60°C for an hour. Store the inserts in a desiccator or in an air tight container until use.

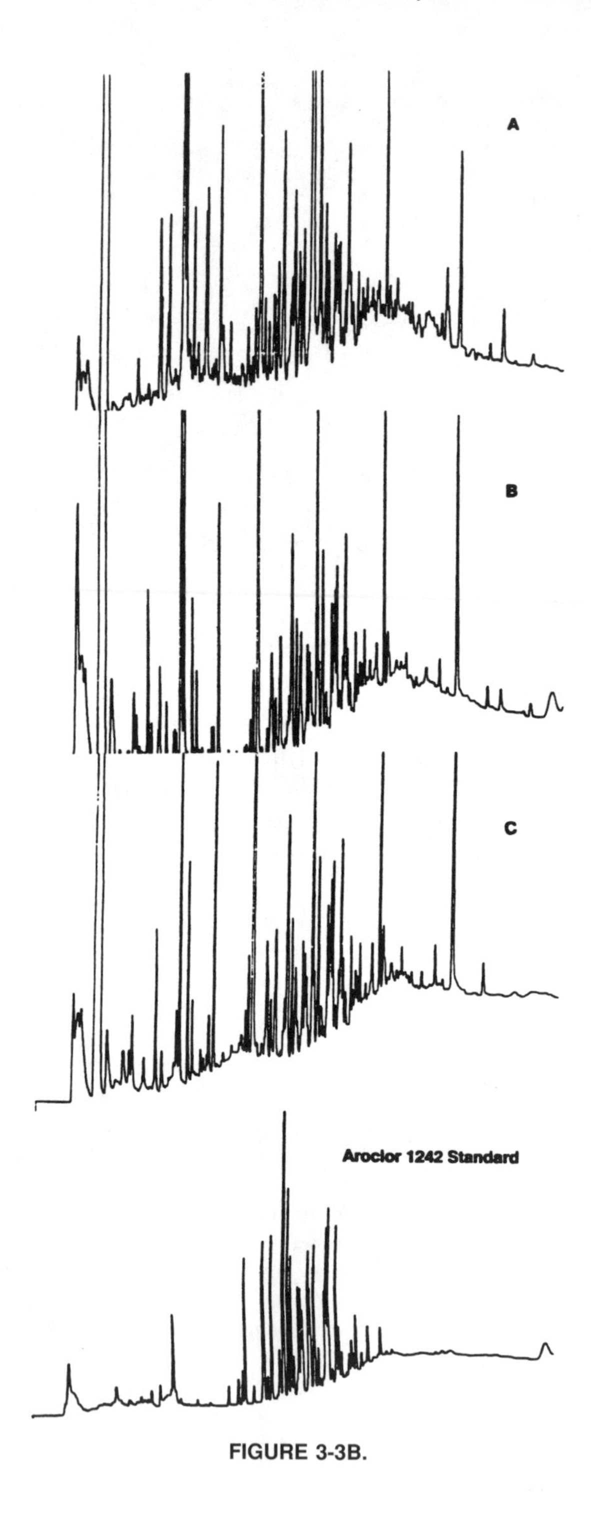

FIGURE 3-3B.

You can deactivate injector liners by coating them with a GC liquid phase

Fresh injector liners will help keep response factors constant and will improve chromatography for acid/base compounds. Alternately, you can coat the injector liner with a liquid phase. This further improves performance, especially for higher boiling components such as PAHs. In this case use the following procedure:

- Clean the injector liner as above and bake out to remove water.
- Prepare 3% w/w solution of a nonpolar liquid phase (such as OV-1, OV-3, OV-7, OV-17, etc.) in toluene.
- Place the inserts into the solution and store until ready to be processed.
- Remove the inserts from the solution and place in a drying oven at 60°C for 2 to 4 h to remove the toluene.

The inserts coated with the liquid phase will last longer than the ones simply deactivated with HMDS/TMCS. The recovery for the PAHs will also improve by 10 to 20%.

Leaking Septa

If the septum is leaking you will lose sample every time you inject. The cause of this is usually an old septum that has been repeatedly punctured by the syringe. The gauge and the shape of the needle also affects septum life (i.e., septa don't last as long when you use large gauge needles with a high bevel angle — the needle will "core" out a piece of the septum each time you inject).

You can check for leaks easily by shutting off the carrier gas

If a capillary column is being used, a simple check for leaking septum can be done by turning off the split gas and septum sweep flow and turning off the carrier gas before it gets to the GC or turning off the gas at the tank. After turning off the gas, monitor the rate of decrease of the column head pressure. If the head pressure falls by more than a few pounds in 20 or 30 sec, a leak is evident. If you notice the head pressure falling, the septum is the logical place to begin your inspection.

Note: *If a packed column is in use, it's not a good idea to remove the septum cap without first bleeding off the head pressure.* If you don't release the head pressure the packing may blow out of the column and into the injector because of the pressure differential. The best procedure is to lower the injector temperature, shut off the carrier gas, and allow the head pressure to drop to zero. Then the septum can be removed without any problems — like burning your fingers and blowing column packing all over the lab.

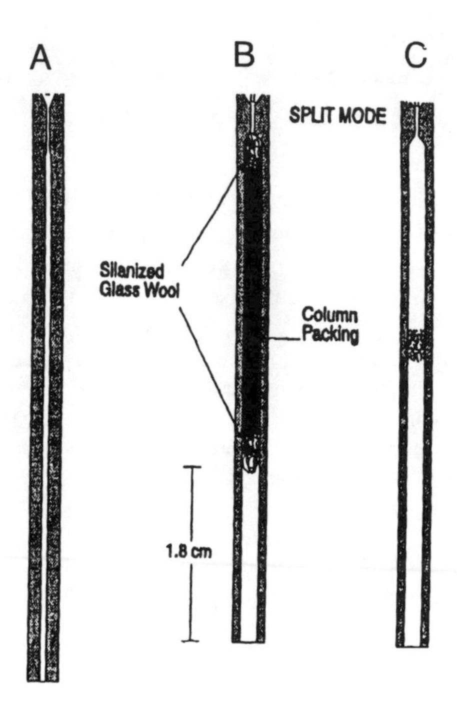

FIGURE 3-4. Examples of injector inserts for split and splitless injections. The relative standard deviation for these inserts is about 2%. If the tubes (B and C) were empty the relative standard deviation would be about 35%.

Leaking Ferrules

You can use the same test as above for checking for leaks due to loose or cracked ferrules. Obviously, if you can't stop a leak by tightening the ferrules you need to replace them. Some injectors (e.g., Varian) use ferrules inside the injector. So, it's important to use the right size ferrule in these cases. If you don't, there can be leaks inside the injector that are very hard to detect and you will lose sample and sensitivity.

Foreign Material in the Injector

If you get foreign material (e.g., graphite) in the injector it will absorb the compounds you are trying to detect. If you get a small piece of graphite in the injector your sensitivity will fall to just about zero — you won't see anything but the solvent when you inject. Graphite ferrules especially have a tendency to break off small pieces and get into the injector. If this happens you will immediately notice that your sensitivity has gone down. If this is the case you need to remove the graphite from the injector. We have used carrier gas (or anything noncombustible) to blow foreign material out of the injector.

Procedure to remove foreign material from the injector

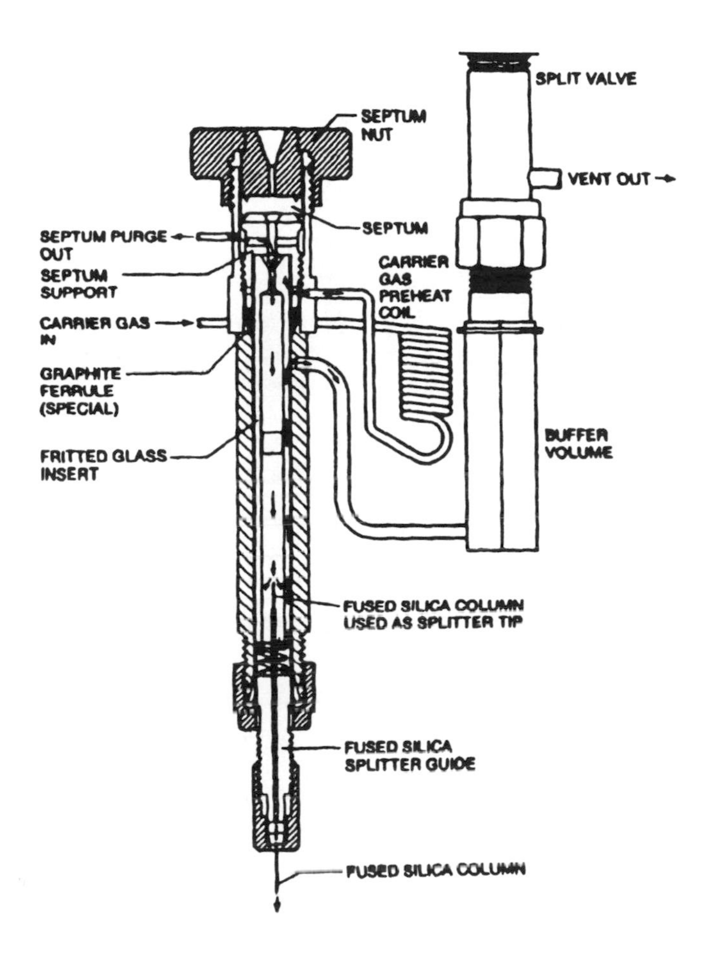

FIGURE 3-5. Diagram of a typical injector.

- Turn the tank pressure at the outlet stage of the regulator to zero and bleed off any residual pressure. If the regulator has an on/off valve after the second stage leave it open.
- Hook up a piece of rubber tubing to the tank. You can use tygon tubing if that's more convenient, but you might get some phthalate background after the injector is cleaned.
- Hold the tube vertically about a foot away and *very slowly* turn up the pressure on the tank until you can just hear the gas. You should be able to feel the gas flow with your finger.
- Insert a long tipped Pasteur pipet into the tube. Be careful that the gas flow is not too high or the pipet will blow out of the tube. You want a nice stream of gas coming out of the pipet.
- Insert the pipet into the open injector and move it to the bottom of the injector. The foreign material will be blown out the top of the open injector.

To avoid getting graphite into the injector we have found that it is good practice to cut off a small piece of the capillary column after you slide the nut and ferrule over

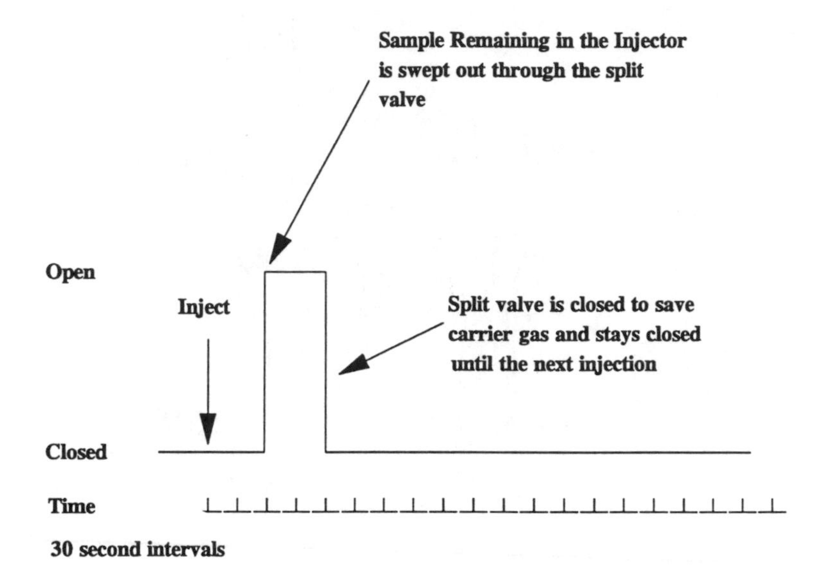

FIGURE 3-6. Split valve timing cycle. The sweep valve is closed for 1 min after injection to allow the sample to deposit on the head of the column.

the end of the column. That way, if there is a small amount of graphite on the end of the column from the ferrule, it will not get into the injector.

Peak Splitting

Sometimes you might notice that a chromatographic peak is split. That is, the same component appears as two peaks that may not be resolved. This is the result of a double injection. Double injections can be caused if the outside of the needle is wet with solution, if the plunger depression is not continuous, or if the pressure on the head of the column changes during the injection.

Leaky septums contribute to peak splitting

The first two problems are easy to correct by wiping the needle and injecting in a continuous motion. The pressure change problem can be tricky. When you make an injection the vaporized sample enters the column and begins to move down its length. If the pressure is lowered at the head of the column some of the material you just injected moves backward because the pressure *in* the column is higher than at the head (injector). This results in the sample being deposited in two places on the column. If the septum leaks only a little during injection you will see split peaks (i.e., a double injection). Peak splitting is not very common but it does happen from time to time. If you observe this you must find out why the head pressure drops and eliminate the cause — usually a leaking septum.

Columns

The type of column you use will affect elution order, chromatographic resolution, overall retention time, and capacity for sample. Although packed columns are still used in a few laboratories, most labs have switched to capillary columns because of the better chromatographic resolution. Our discussion will, therefore, only deal with capillary columns.

The elution order of the components is primarily affected by the liquid phase used to coat the capillary. Capillary columns are made of fused silica (quartz) glass and are deactivated and coated with the liquid phase. Bonded phase columns use a liquid phase that polymerized when the column was heated the first time. The liquid phase is then "bonded" to the column. Thus, bonded phase columns tend to be more stable over longer periods. Liquid phases may be nonpolar (e.g., OV-1, which is a methyl silicone), slightly polar (e.g., OV-17, which is a methyl phenyl silicone), or very polar (e.g., OV-275, which is a dicyanoallyl silicone). The elution order of the components will depend on the liquid phase being used. Most environmental analyses are performed with either a nonpolar or slightly polar liquid phase. Capillary columns with a bonded phase such as DB-1 or DB-5 are fairly common. The DB prefix indicates a bonded phase column and the number indicates which liquid phase is being used. The 1 in DB-1 means the column behaves like OV-1; the 5 of DB-5 is similar to OV-17, etc. As you can see, the numbers don't always match; however, the higher the number the more polar the phase.

The approximate number of plates per meter of the column is given by the reciprical of the internal diameter

The internal diameter (r) of the column will have a pronounced affect on column resolution. The smaller the value of r the more theoretical plates per meter. In fact, you can estimate the maximum number of plates per meter by dividing the value of r into 1. For example, a 0.2 mm inner diameter column can have up to 1/0.2 mm or 5000 plates/m. A 0.75 mm inner diameter column can have about 1333 plates/m. Thus, all things being equal, the narrower bore columns will provide better separations than the wider bore columns. So, you should always use narrower bore columns right? Well, things aren't always equal and wide bore columns do have advantages. It's safe to say that if you want maximum resolution you should use the narrowest bore column.

Advantages of wider bore columns

Wider bore columns can accommodate a larger film thickness than narrow bore columns. The film thickness influences how much material can be injected or the "capacity" of the column. For example, columns with a 0.25 mm inner diameter that is coated with a 0.25-µm film of the liquid phase will begin to saturate if you inject more than about 100 ng per component. A wider bore column with a film thickness of 0.5 to 1.5 µm won't begin to saturate until you inject about 1 µg per component or possibly more.

Check peak shapes for saturation

Saturation means that the liquid phase cannot dissolve or equilibrate all of the analyte. When this happens the chromatographic resolution of the system degrades. You can tell when you saturate a column by looking at the resulting peak shapes. The peak shape will be asymmetric (see Figure 3-7), with the peak looking like a right triangle. If you see peak shapes like this you have overloaded the column and saturated the liquid phase. If this happens, you should consider diluting the sample and reanalyzing.

Film thickness and the inner diameter of the column affect capacity. The film thickness alone is one indication that the column has a higher capacity. The inner diameter of the column is another. If you consider two columns — one a 0.53-mm inner diameter column with a film thickness of 1.5 µm and the other a 0.75-mm inner diameter column with a film thickness of 1.5 µm, the 0.75 mm column has about 30% more liquid phase, which means it has about 30% more capacity. Thus, film thickness, column inner diameter, and column length determine a column's resolution and capacity. Environmental samples often have a wide variability in concentration and, thus, the wider bore columns with greater film thickness and higher capacity are generally more useful. In effect you trade resolution for capacity.

Wide bore columns with high film thicknesses are especially useful for the analysis of volatile compounds. Thicker film columns trap volatile components more efficiently at higher temperatures than thinner film columns. You can tolerate higher oven temperatures when you flush the compounds from the trap with thicker film columns. This is a distinct advantage over thinner film columns. For example, narrow bore columns cannot trap the volatile components effectively unless you cryofocus (i.e., you have to cool the head of the column to below 0°C). Now, it is possible to use a 0.25-mm inner diameter column for volatiles, but it is difficult to effectively trap the gases (i.e., the first four compounds of interest) on the column. A better option is to use a wide bore column — either the 0.53 or 0.75 mm inner diameter columns with a 1.5- to 3-µm film thickness. In order to achieve the same resolution as a narrow bore column, however, you will likely have to increase the column length from 30 to 60 m. Wide bore columns 100 m in length or more are not unheard of for this application.

The wide bore columns can be used for the gases at subambient temperatures (e.g., 10°C) and cryofocusing is not necessary. The thicker film of the wide bore allows the column to operate at a higher temperature and still effectively trap the volatile compounds that are flushed from the analytical trap. The thicker the film the better the column will act to trap and focus the volatile compounds. Therefore, thick

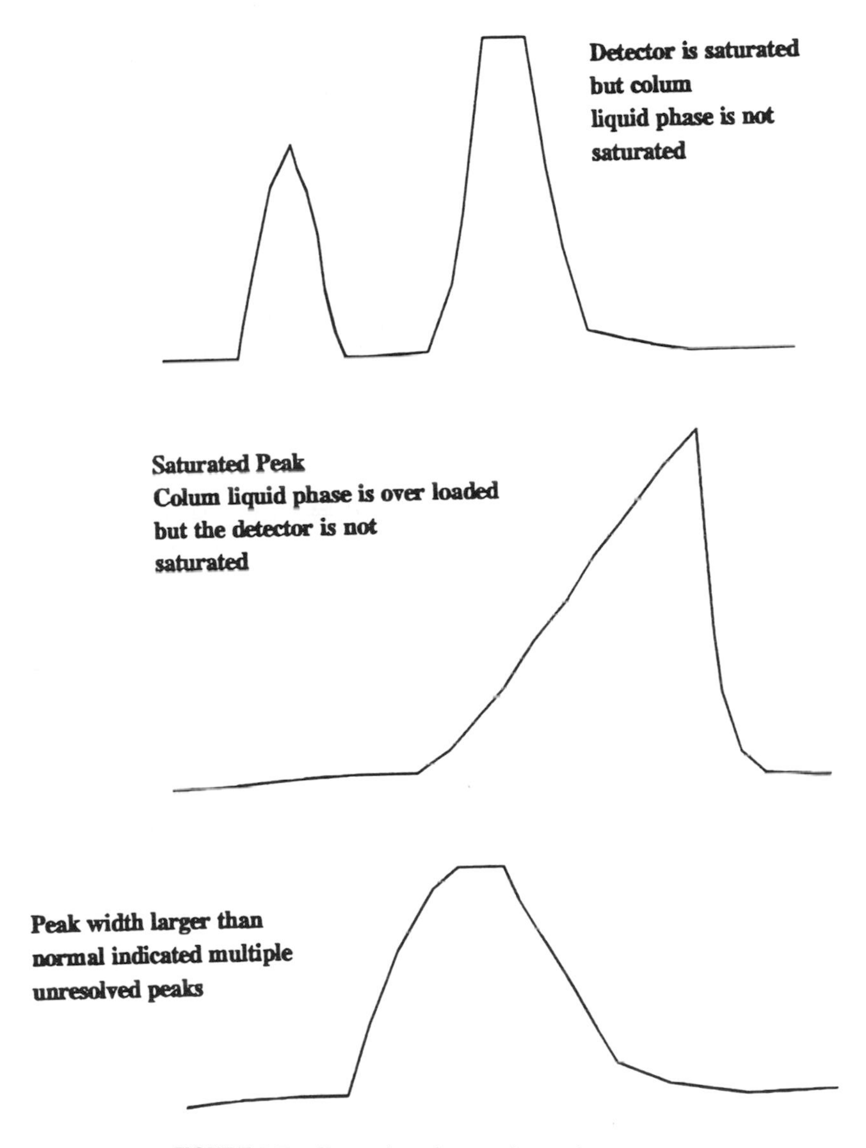

FIGURE 3-7. Examples of saturation and asymetry.

film columns have advantages in capacity and their ability to trap volatile compounds. Because of these two features many laboratories use wide bore columns for volatile analyses and analyses where the concentrations of the components can vary by several orders of magnitude.

The best of both worlds. Use a short piece of Wide bore column coupled to a narrow bore.

We believe that narrow bore columns are best for most environmental analyses since they provide the best resolution. The apparent drawback to cryofocusing can be overcome by using a short (e.g., 0.5 m) piece of a thick film wide bore column in conjunction with a 0.25 mm column. Basically all you need to do is to hook up a piece of 0.53- or 0.75-mm inner diameter column to the injector and then butt connect this short column to a 30- m, 0.25-mm inner diameter column. You will get the benefit of subambient trapping in the GC oven while still getting good resolution from the narrow bore column. Also, by using a narrow bore column you can avoid additional steps and problems when you go to GC/MS. This will be explained in the following sections. The only drawback is the limited capacity of the narrow bore column.

Acid/Base Behavior

Column failure may cause low sensitivity for polar compounds

If you have low sensitivity or peak tailing for polar compounds then the injector and/or column might be more active toward acids or bases (i.e., polar compounds). Generally, this problem is caused in the injector liner; however, the column may also contribute. The best way to check this is to inject a test mix that comes with the column and see if you're losing one or the other of these types of compounds. The test mix will contain one or more acidic (i.e., phenols) and basic (amines) compounds. If you are losing acidic or basic compounds the first thing is to remove the glass injector liner and replace it with a clean, deactivated (silanized or coated) liner. If the problem persists it is probably due to the column. In this case try cutting off about the first 30 cm of the column and reconnecting it to the injector. Reanalyze the test mix. If this fails you will probably need to replace the column. Columns typically cost $400 to $500. Considering that you may lose that much in revenue by spending 2 or 3 h evaluating the column, it may be less expensive to simply replace the column before cutting the first part of the column off.

Loss of High End

If the injector, column oven, or GC interface are not hot enough the less volatile compounds in the sample will not be transferred to the detector. Manufacturers recommend that these system components must reach a temperature slightly above the temperature where the component elutes from the column. Thus, if the highest boiling component elutes at 250°C then the injector and interface should be maintained at 260 to 270°C. If any component of the system is cooler than the elution temperature of the highest boiling components these components will elute as broad peaks or may not elute at all. This condition may also be caused by having too fast an injection.

Injection Speed

The speed of the injection has an effect on the resulting chromatogram, which is probably due to nonequilibrium conditions in the injector. Very fast injections will display a reduced high end. For example, the response for PAHs can be lower if you make very fast injections. This is due to a differential vaporization during the injection. A fast injection does not ensure that the high boiling compounds are vaporized to the same extent as the lower boiling components. This means that the response for the higher boiling components will be decreased. One way around this is to increase the injection time by slowing the injection speed.

Decrease high end discrimination by slowing injection speed

Some autosamplers are capable of making very fast injections and subsequently may have this high end discrimination. The HP-7673 autoinjector can make very fast injections and high end discrimination is possible. One way to overcome this is to use the longer "on-column" injection speed. This requires that the injector be raised about 3/4 in. above the GC to accommodate the longer injection stroke used during on-column injections. This can be easily accomplished by mounting the autosampler on a 3/4-in. plywood insert between the GC and autosampler. This permits using the slower on-column injection speed while keeping the injection depth the same.

While on the subject of autosamplers, we have noticed that the heat from the column oven can evaporate samples from their vials after the sample vial has been punctured. For example, if you make one injection from a given vial and then run a series of samples, and then make another injection from one of the vials you will notice that the concentration reported from the second injection will be higher, indicating that the sample is more concentrated. This happens since the solvent in the vial is being evaporated while the other samples are being analyzed. This is not really a major problem unless you have to run a series of standards and then must confirm by using the same standard. Unfortunately, this may be the situation you are in with pesticide analyses. Many EPA methods require the calibration be verified several times during analysis. If you use the same vial, the results will show that the calibration is off. The reality is that the standard is getting more concentrated. One way to avoid this is to use multiple standard vials and only use them once. Another way to get around this is to provide cooling to the autosampler via an external fan. Many labs have used both approaches and found them equally satisfactory. If you use the external fan approach all you need to do is to mount a small muffin fan on the autoinjector above the sample carriage so that the air blows down on the autoinjector and the sample vials. This cool air will minimize evaporation and allow several injections to be made from the same vial.

Autosamplers and Sample Carry-Over

It is possible to get "carry-over" or cross contamination between a highly contaminated sample and a relatively "clean" sample when using an autosampler. If the syringe is not adequately cleaned or if some residue remains in the injector, on the injection liner, or on the septum you will see the same chromatogram as the previous injection. The chromatogram will be attenuated; however, it will look almost identical to the previous injection. The best way to avoid this situation is to take care in cleaning the syringe and any lines used to fill the syringe between injections.

Clear up carry-over by injecting blanks

The pump-type autoinjectors (i.e., the ones that fill the syringe by pumping the syringe in the sample) are much less prone to sample carry-over than the continuous flow-type autoinjectors (i.e., the ones that require a constant flow of sample through the syringe). Regardless of the type of autosampler you have, you may still notice carry-over between highly concentrated samples and samples with lower concentrations. If a sample is obviously highly contaminated (by visual inspection or it exhibits an odor) analyze it at the end of the sample batch. You can check for sample carry-over by injecting a blank between samples. If you have a carry-over problem the blank will exhibit the same type of chromatogram (attenuated) as the previous sample. Usually by the second or third injection of the blank (and subsequent GC program) the carry-over will be gone and you can resume analysis.

If the carry-over does not go away after three injections you will probably have to replace the injector liner and the septum. (Septa can absorb compounds and every time you pierce the septum you will be introducing those compounds into the injector.) Replacement of the injector liner and septum will correct most carry-over problems, assuming the problem is not in the carrier line or in the autosampler.

We had one occasion, however, when we could not solve our carry-over problem by these means. In fact, the only way to get the background back to a reasonable level was to replace the GC column. In this particular case we injected a sample that contained a high concentration of pentafluorobenzylbromide (PFBB) derivatizing reagent. We tried replacing the liner and the septum — to no avail. We even cooled the injector and washed it out with solvent. This didn't work either. As it turned out the gas chromatographic column (i.e., the liquid phase within the column) had absorbed enough of the PFBB that we could not use the column for low level work. The PFBB would elute continuously from the column at temperatures above 150°C and the baseline would shift off scale. Therefore, if all else fails, replace the column.

Gases

Gases used for gas chromatography fall into three categories, which are

- Carrier gas
- Detector gas(es), including make up gas, if necessary
- Utility gas

As the name implies, the carrier gas is used to transport (i.e., "carry") the sample through the chromatographic column to the detector. The choice of carrier gas will depend to some extent on the detector being used and on the chromatographic resolution required for the separation being performed.

Carrier Gas and Chromatographic Resolution

Hydrogen used as a carrier gas offers the best chromatic resolution and price

Different carrier gases will provide different degrees of chromatographic separation with respect to the carrier flow rate. Without getting into a detailed explanation, hydrogen is the carrier gas that will generally provide the best chromatographic resolution over a wide range of flow rates. This means that you can run with very high flow rates and maintain the best chromatographic resolution. The fast flow rate means that the analysis time is shorter. You can double the flow rate with hydrogen while maintaining about the same resolution. Double the flow rate means half the analysis time. Most chromatographers prefer hydrogen for two reasons — it's relatively cheap (about one third the cost of helium) and it provides for the most efficient (i.e., fast analysis time) separations. Normally, carrier gases are discussed in terms of the HETP (height equivalent of a theoretical plate). If you think of a column as being composed of individual distillation "plates" then obviously the smaller the height of each plate, the more plates can occupy a given length of column. Thus, at a given flow rate, if the HETP for hydrogen is 0.5 mm and the HETP for nitrogen is 1.0 mm, then the separation using hydrogen will provide double the number of plates for a given column length. As a rule of thumb, hydrogen is normally the best choice, followed by helium, then nitrogen for a carrier gas. A tank of helium costs between two and four times more than a tank of hydrogen or nitrogen of comparable purity. The main point here is that you can use hydrogen at very high flow rates and still maintain resolution.

Any disadvantages of hydrogen are outweighed by using it safely

Many people think that hydrogen presents an unacceptable risk of explosion in the laboratory if it is used as a carrier gas. Fortunately, hydrogen diffuses very rapidly, so the danger from explosion is fairly remote unless a massive leak occurs. For hydrogen to detonate in air the concentration must be in the percent range. If you are using a wide bore capillary column the flow rate will be 15 to 20 ml/min. If the column breaks off in the oven the hydrogen should not reach a high enough concentration to cause a problem. As long as reasonable precautions are taken against developing a serious leak (i.e., liters per minute) we feel hydrogen's advantages far outweigh any potential disadvantage.

Carrier Gas and Detectors

As indicated above, some detectors will not work very well with certain carrier gases. Fortunately, all detectors we are aware of will work well using either hydrogen or helium as a carrier gas. A few labs complain about using hydrogen as a carrier gas with electron capture detectors (ECDs). We have not experienced this problem and use hydrogen as a carrier for all of our gas chromatographic separations. ECDs, however, require a make-up gas and we use helium or argon methane 95:5 for the make-up. We have not tried to use hydrogen as a make-up for ECD and perhaps this is the basis for the negative comments regarding hydrogen and ECDs. Obviously, gases that interfere with a given detector cannot be used as a carrier (e.g., nitrogen cannot be used for a thermionic [N/P] detector).

Carrier Gas Flow Control

Set the gas tank pressure first

The flow of carrier gas through the GC column is controlled by either a pressure regulator or a flow controller. A pressure regulator is most often used with capillary columns. In this case you set the inlet pressure at the gas tank higher than the pressure you want at the head of the column. You then use the pressure regulator at the GC to adjust the column head pressure. For example, if you want the head pressure of the column to be at 30 psi, then you would set the pressure at the tank (on the second stage of the regulator) to 40 to 60 psi. You would then use the regulator at the GC to set the inlet pressure to 30 psi. If either the regulator at the gas tank or the regulator at the GC is defective you will get variable retention behavior. Pressure regulators can go bad. If you notice erratic retention behavior you should replace the tank regulator to see if the problem clears up.

If you are using an instrument with a flow controller for capillary work you might consider setting the pressure at the tank to the desired head pressure and setting the flow controller all the way on (i.e., to the highest flow rate). The flow controller will try to reach a high flow by increasing the inlet pressure to the maximum tank pressure. Since the flow controller will not be able to reach the high flow, the head pressure will remain constant at the tank (maximum) pressure, which you set at the second stage of the pressure regulator. For example, if you want 30 psi on the column, set the tank to deliver 30 psi. With the flow controller all the way open the pressure at the column will remain at 30 psi. Older instruments generally use flow controllers for setting column flow rate. If you have an older GC (e.g., pre-1980) that you want to use for capillary work you may consider using this approach.

Regardless of how you set up the system, the inlet pressure should remain constant during the chromatographic cycle. The flow through the column, however, will vary depending on the oven temperature. As the GC oven heats up the viscosity of the carrier gas will change. This means that for a given head pressure the flow of gas through the column will change. For example, a flow rate of 1 ml/min of He through a capillary at 50°C may only be 0.75 ml/min at 250°C. This is of little consequence as long as the conditions remain the same from run to run.

With capillary columns it is more common to talk in terms of "linear velocity" of the carrier rather than absolute flow rate. Helium gas goes through a minimum HETP (i.e., the best chromatographic resolution) at a linear velocity of about 30 cm/sec. This would correspond to a flow rate of about 0.9 ml/min for a 0.25 mm inner diameter × 30 m column. **Note:** *At 30 cm/sec the column volume is replaced every 100 sec.* The column volume is 3000 cm $\times \pi \times$ $(0.025/2)^2$ cm^2 or 1.47 cm^3. As mentioned above, you can easily double this flow for hydrogen and still maintain the chromatographic resolution.

Detectors

The most commonly used detectors are the flame ionization detector (FID) and the electron capture detector (ECD). Other detectors frequently include the electrolytic conductivity, thermionic, flame photometric, and photoionization detectors.

The Flame Ionization Detector

The sensivity of an FID is similar to that of a Mass Spectrometer

FID response is poor only for non-burning and halogen compounds

The FID is a nonspecific detector. That is, it responds to almost all compounds that contain carbon that may be oxidized (i.e., converted to carbon dioxide) in a flame. The FID is generally recognized as a "universal detector" since it responds to most organic compounds. The only compounds that it does not detect are those that do not burn (i.e., not readily oxidized). For example, water and carbon dioxide do not give a signal using an FID since they are not oxidized in a hydrogen flame, because they are already in their highest oxidation states. Furthermore, compounds that contain halogens do not give as large a response as compounds not containing halogen. Halogen atoms are good scavengers for free electrons. Since a hydrogen flame is a chemical reaction that relies on free electrons (radicals) and ions, the response to halogenated compounds is usually smaller compared to molecules the same size that do not contain halogen. The electron capture detector and the electrolytic conductivity detectors are more specific than the FID. Their primary advantage is their sensitivity to specific compound classes (e.g., halogens).

The FID is a general-purpose detector that finds wide use in the lab (see Figure 3-8). It can be used for sample screening to estimate how much a sample must be diluted or concentrated to ensure that the determination is made in the calibrated range for the analysis. It is also the detector of choice for certain types of compounds, e.g., hydrocarbons. The FID is about as sensitive as a mass spectrometer. That is, you

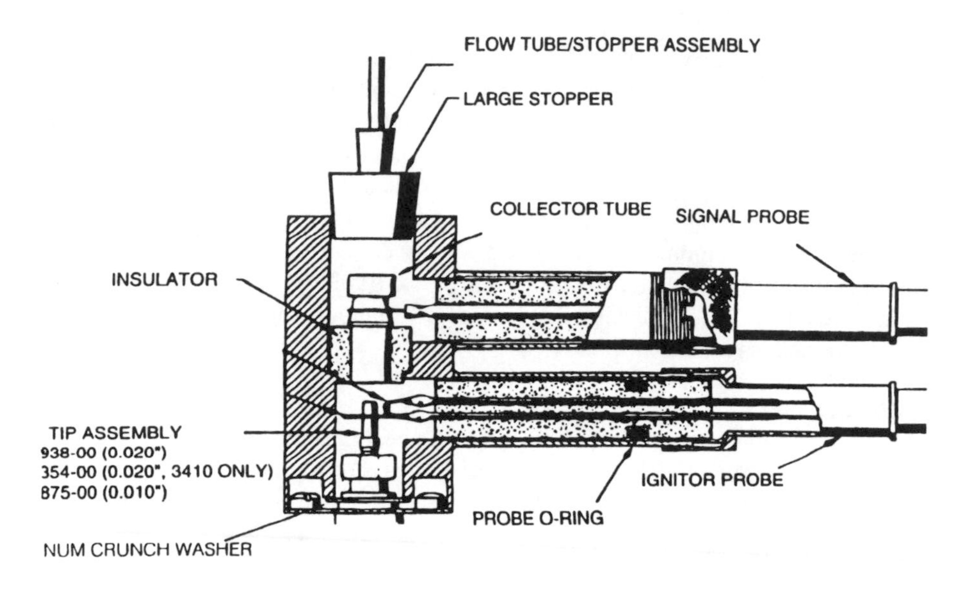

FIGURE 3-8. Diagram of a typical flame ionization detector.

need to inject about 10 ng per component to get a useful signal. Therefore, if you are calculating whether to concentrate or dilute a sample, try to make your estimate based on delivering 10 ng "on-column".

Observe condensation to determine if flame is lit

The FID is a very simple detector. The detector burns hydrogen in the presence of excess air to form water. A hydrogen flame is not visible to the naked eye so one usually uses an indirect method to "observe" the flame. Typically, we use a mirror, piece of glass, bottom of a beaker, wrench, etc., to observe whether the detector is lit. Anything with a smooth surface that will condense the water vapor resulting from the flame can be used to determine if the flame is lit. You verify that the flame is lit by holding the mirror or piece of glass over the detector chimney and observing the water vapor condense. If there's no condensation the flame is out.

Decrease air flow if the flame won't stay lit

Typically, the hydrogen flow rate is about 30 ml/min, which is about 10 times less than the flow for air. If you ever have problems lighting the flame, or getting it to stay lit, the most common problem is with the air flow rate. If the air flow is too high the flame will not light or it will go out after it is lit. If this is the case, the best method to get the flame to stay lit is to reduce the air flow in increments until the detector lights. The air flow can be decreased to 200 ml/min or even 150 ml/min to get the detector to light.

Another reason the flame detector may not light is a defective igniter. The igniter is normally a glow plug. Make sure you see the plug glowing. If the plug does not glow it may be burned out. In this case replace the plug and try to relight the detector.

Preparation for GC-FID

Setting up a gas chromatograph for an FID analysis usually requires the following steps:

- Install the column in the gas chromatograph. Each chromatograph has recommended distances to use when installing capillary columns. That is, there are specific distances that the capillary should extend into the injector body and into the detector. You should consult the information provided with your GC as to the recommended distances. If you do not have this information you can do either of two things — call the manufacturer to find out the recommended distances or compute them. If you want to compute the distances measure the length of the injector from the top of the septum nut to the bottom of the fitting (nut) inside the GC oven. Measure the length of the syringe needle you will use. Subtract the length of the needle from the overall injector length. Subtract an additional 1 to 2 cm from the remaining length. Measure this distance on the column and mark it with white-out. This is the correct distance to insert the column into the injector if you are using a splitless-type injection insert.
- Turn on the carrier gas and measure the flow. You can do this before the column is attached to the detector. Typical flow for a 0.25 or 0.32 mm inner diameter capillary column is about 1 ml/min at a head pressure of about 10 psi. Larger bore columns, (e.g., 0.53 or 0.75 mm) will run between 5 and 15 ml/min at about 5 psi. Remember, if you are using hydrogen you can use higher flow rates and still get about the same separation efficiency. See the discussion on carrier gases.
- Turn on the hydrogen gas to the detector (make sure the column is installed) and measure the flow at the output of the detector. If the column flow rate is around 1 ml/min, you can ignore this contribution to the hydrogen flow. Adjust the flow to about 30 ml/min or to whatever the manufacturer recommends. (It should be around 30 ml/min.)
- Turn off the hydrogen and turn on the air. Adjust the flow to about 300 ml/min. Turn the hydrogen back on.
- Try to light the detector — make sure the detector heater is on and the detector is at temperature. You will hear a "pop" when the detector lights. To see if the detector is lit, use a piece of glass, mirror, wrench, etc., and hold it over the chimney of the detector. If you see water vapor condense on the piece, the detector is lit.
- If the detector does not light, or stay lit, lower the air flow. You don't have to measure it, just lower it until the flame lights and stays lit. You can then increase the air flow to 250 or 300 ml/min.

FIDs do get dirty and require cleaning from time to time. One symptom of a dirty FID is shot noise. Figure 3-9 shows an example of a chromatogram from a dirty FID.

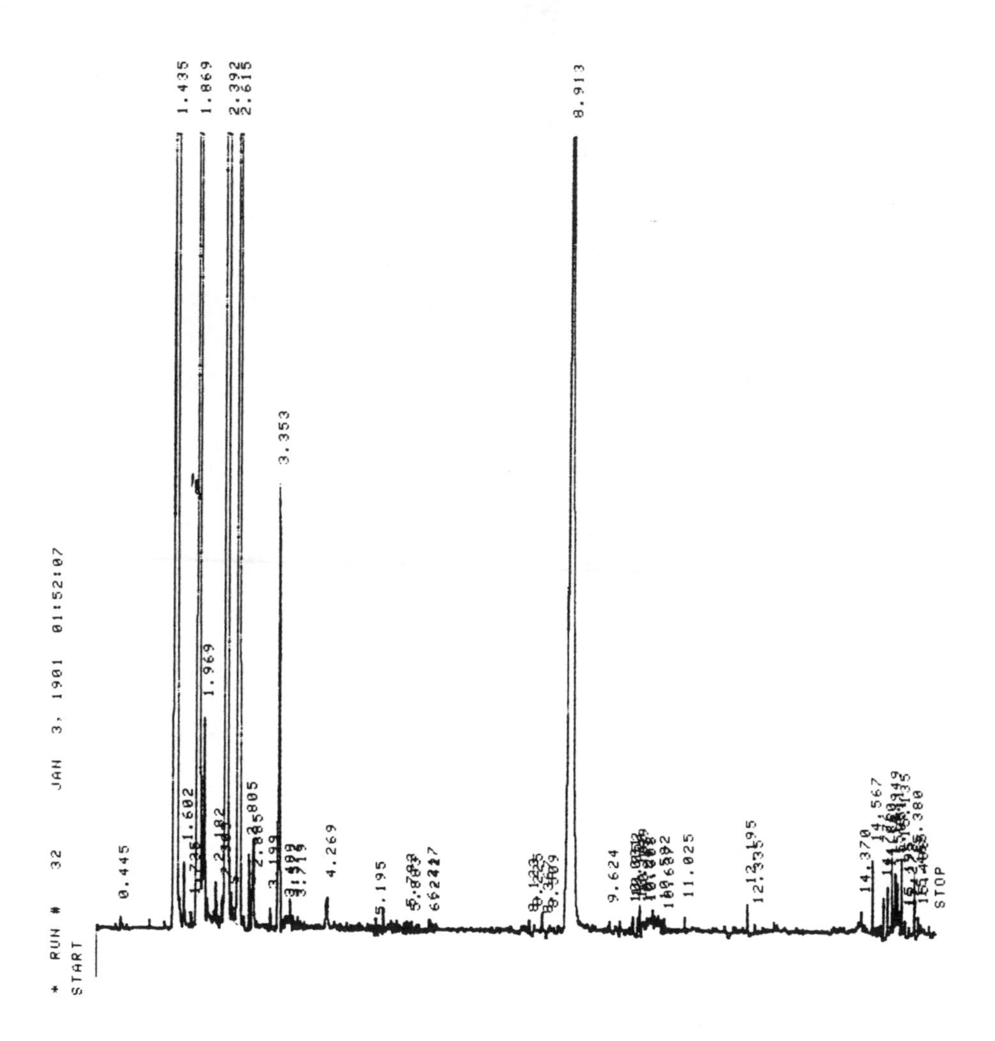

FIGURE 3-9. Example of a chromatogram from a dirty FID.

The chromatogram displays several spikes in the baseline and on the chromatographic peaks themselves. The spikes are due to a sudden change in the current at the collector. This is caused by bits of material that plated onto the collector breaking off. This problem can be remedied by cleaning the FID.

Cleaning an FID is a very simple procedure. After the power to the instrument is shut off and the detector cooled down, the collector is removed. The collector is cleaned by passing a stiff brush through it. We have also found that pipe cleaners impregnated with wire also do a good job at removing deposits from the collector of an FID. After all of the material on the collector has been removed, reinstall the collector.

Electron Capture Detector

The electron capture detector responds selectively to electron-absorbing compounds. An electron source (e.g., ^{63}Ni) is used to generate a standing current of thermal electrons (see Figure 3-10). This standing current is converted into a baseline signal. When an electron-absorbing compound enters the detector the standing current is lowered, which gives rise to the signal. The electron capture detector normally uses a make-up gas to sweep the detector. The make-up gas may be an inert gas, such as nitrogen or, commonly, a mixture of 5% methane in argon. The argon-methane mixture enhances the sensitivity of the detector. The flow rate for the make-up gas is typically about 30 ml/min and the flow from the column will be 1 to 10 ml/min, depending on the type of column being used. Electron capture detectors rarely need to be cleaned. In fact, I couldn't find an example of a chromatogram from a dirty ECD

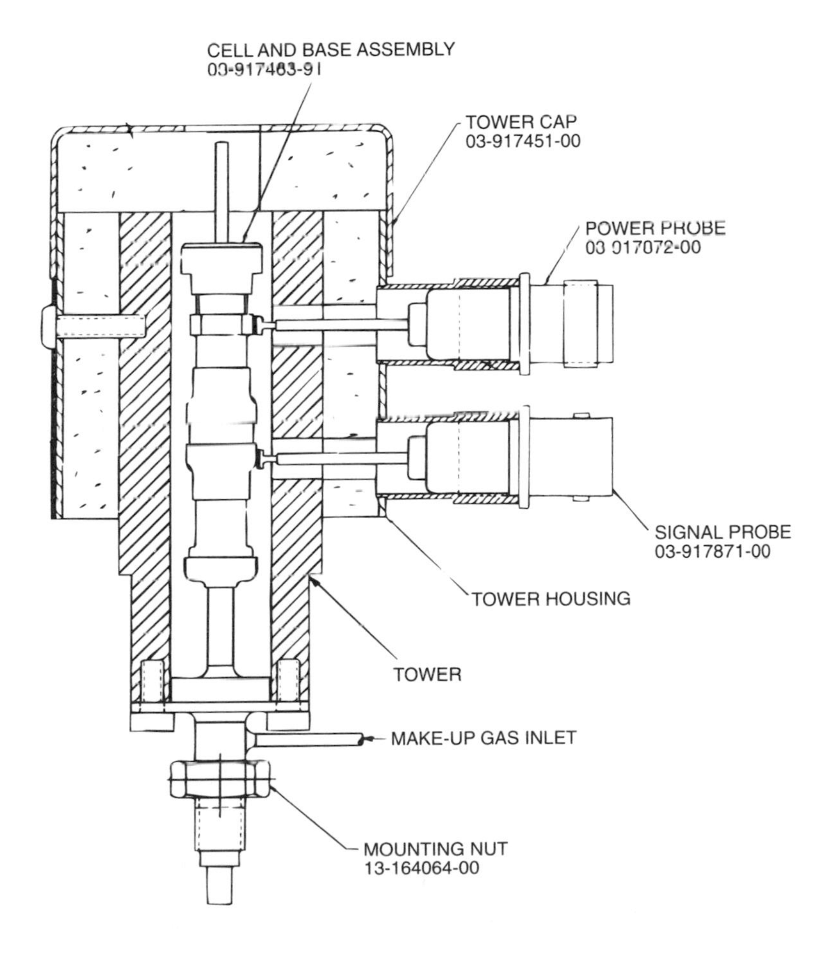

FIGURE 3-10. Example diagram of an electron capture detector.

when writing this book. An ECD that loses sensitivity over time (several months) is probably getting dirty. That is, the standing current from the ^{63}Ni is lower. If you suspect your ECD is dirty consult your user manual for instructions or call the manufacturer.

Photoionization Detectors

Photoionization detectors (PID) use ultraviolet light to ionize molecules with electron affinities less than about 11 eV. The ionization energy of some PIDs can be "tuned" — the energy can be lowered so the detector only responds to specific compound classes. Thus, the PID can be configured to respond to compounds with some type of substitution or functional group. In particular the PID is an especially good detector for PAHs.

Keep the PID window clean

There are a couple of problems you may encounter when using a PID. The lamp can burn out quite rapidly with very little warning. It is a good idea to keep a spare lamp on hand for such an eventuality. Another problem may result from a build up of residue on the window of the detector. See Figure 3-11. Compounds can "plate-out" on the optical window of the detector and lower the efficiency of the detector by limiting the amount of light entering the detector cell. One way to avoid this is to use a make-up gas flow rate of about 20 ml/min. Some detectors use a sweep gas flow to keep the window clean. The detector temperature should be about 250°C to minimize sample condensation in the detector or on the window. As the window gets dirty you will notice a slow decrease in sensitivity over time. The way to correct this is to disassemble the detector and carefully clean the window per manufacturer's specs.

Electrolytic Conductivity Detectors

When using an electrolytic conductivity detector the effluent from the gas chromatographic column is combusted to form HX where X is Cl, Br, or I, the HX is transferred to a conductivity cell for measurement, and the residue (i.e., the anions X^-) are absorbed on an exchange resin. Therefore, problems with an electrolytic conductivity detector emanate either from the combustion process, the conductivity cell, or the resin and pump system.

Compound and Mixture Identification and Quantification

In gas chromatography the identification of compounds is done by use of specific detectors and comparing retention behavior on one or more columns (liquid phases). If the retention time, or relative retention time to an internal standard, of an unknown matches the retention behavior of a standard, then the unknown is tentatively identified. This identification is then "confirmed" by using a second column (liquid

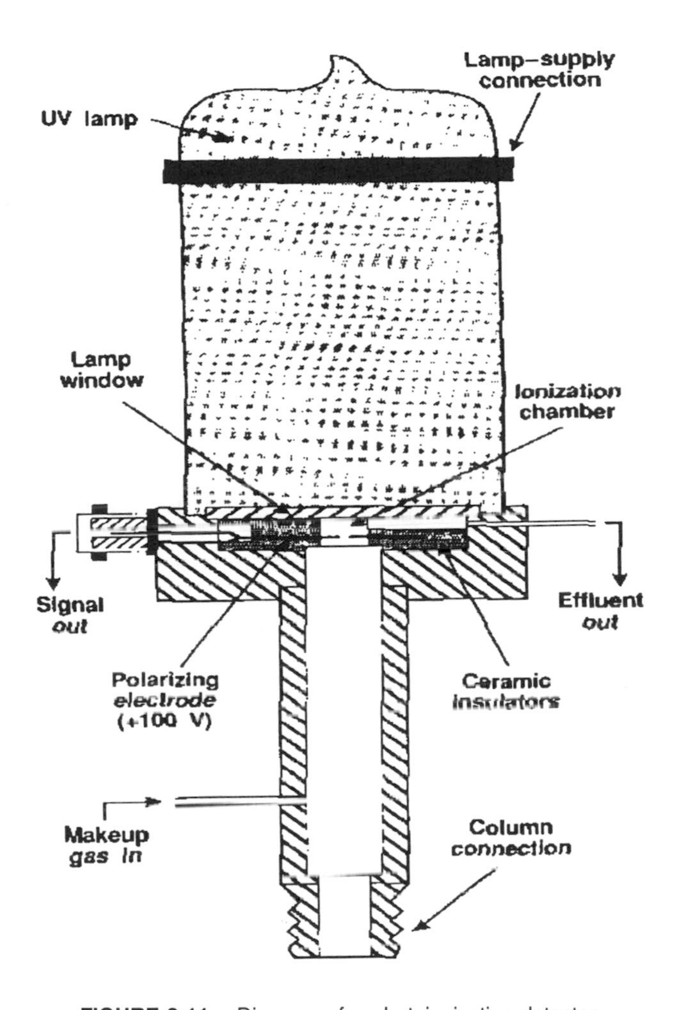

FIGURE 3-11. Diagram of a photoionization detector.

phase) or by using mass spectrometry. Thus, a component of a sample must give the same retention time (within ± some limit) as the standard on two columns or be confirmed with mass spectrometry. There is always a possibility for error in making an identification or missing a component and these errors are referred to as false positives and negatives.

A functional view of false + and –'s

The likelihood of a false positive (saying a compound is present when it is not) or a false negative (saying a compound is not present when it is) is related to the specificity of the detector, gas chromatographic columns, and above all, the concentration of the analyte. Rather than speak of false positives and negatives as defined above it is more useful, especially for environmental analyses, to consider a false-positive or -negative condition as a function of the analytical technique and concentration.

Reporting a false positive or negative will be more likely for components of a sample that are close in concentration to the detection limit (i.e., within a factor of 10 of that limit). Some laboratories and regulatory agencies have adopted a two-level approach. The first level is to call the compound present but there may not be enough of the compound in the sample to accurately quantitate. This means that the sample contains the compound of interest but the concentration is below a "practical quantitation limit (PQL)". The Environmental Protection Agency (EPA) generally defines the detection limit as the lowest concentration that can be correctly identified in a sample 99 times out of 100. That means that at the detection limit the laboratory may give a 1% false-positive response. There are a couple of ways to establish a detection limit for a particular method. Most labs use the following procedure:

- Spike a sample so the concentration will be close to, but slightly above, what you believe the detection limit of the method is.
- Analyze the sample a number of times (usually seven) and compute the standard deviation for the measurement. For example, suppose you spike a sample at 10 ppb and you find the standard deviation at 10 ppb to be 2 ppb. Then the detection limit of the methods is about 3 σ (i.e., the 99% confidence level that the answer is no zero) or 6 ppb.

This approach works fine for clean samples (i.e., samples without interferences); however, it is certainly not the case with routine samples. Therefore we have invented the following rule:

> The detection limits reported by any laboratory are incorrect. In fact, they usually are not even close to the detection limit for an individual sample.

Identification of mixtures (e.g., pesticides) is more complicated than identification of individual compounds, since it requires determining whether the pattern displayed by the mixture in the sample is the same as the pattern of the mixture in the standard. Identification of arochlors (a PCB mixture) represents this type of problem. Most labs have had to establish a series of peak area or height ratios to determine whether a given arochlor is present. The problem with this approach is that arochlor patterns change with time (i.e., the ratios of the individual components change over time as the material ages in the environment). Therefore, you may have an arochlor present in the sample but you might not identify it correctly because the ratios of its components have changed. For example, it may be difficult to distinguish between arochlor 1254 (biphenyl with an average 54% chlorine) and arochlor 1242 (biphenyl with an average 42% chlorine), especially after the 1242 has weathered.

All Arochlor lots are different. Therefore, arochlor standards made from different lots are different.

This situation is further complicated depending on the standards being used. It is not uncommon to find arochlor standards made from different lots of arochlors which are different in composition. That is, a standard from one company may not match that of another. This means that one lab may call a given arochlor present based on their standard while another lab may not identify the arochlor based on a different standard. Unfortunately, the EPA has not published an objective criterion to judge whether an arochlor is present. Each lab has had to develop its own criteria. Furthermore, there is no simple solution to this problem. The best you can do is to use standards from one supplier and develop your own identification criteria (i.e., peak ratios) based on peaks on either ends of the chromatogram. Peaks on one of the ends will have the least amount of overlap with the other arochlors and therefore should provide the best identification criteria. For example, arochlor 1254 will have higher proportions of hexa and hepta chlorinated biphenyls than arochlor 1242. Thus, the hexa and hepta ratios will be good indicators of arochlor 1254 *vis-à-vis* 1242. Figure 3-12 provides an example.

Gas Chromatography-Mass Spectrometry

The second category of detector for gas chromatography is mass spectrometers. A mass spectrometer is a universal detector since it responds to most compounds. The mass spectrometer is highly specific in addition to being universal. Compound identification is accomplished both by using retention time (i.e., the traditional method used in gas chromatography) and the mass spectra of the compound. The mass spectra of a compound are fairly unique to that compound. The only compounds

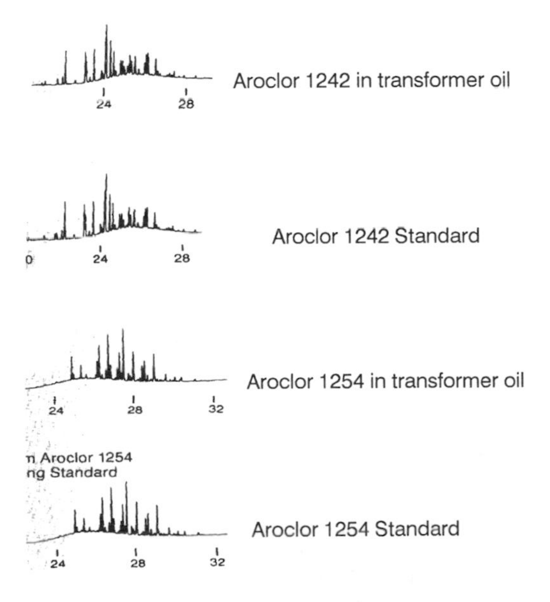

FIGURE 3-12. Comparison of arochlor 1242 and 1254.

that are hard to identify absolutely using mass spectrometry are isomers. For example, it is almost impossible to identify *ortho-*, *meta-*, and *para*-dichlorobenzene isomers based on their mass spectra alone. In this case we rely on the GC retention time to identify which of these isomers are present in the sample.

There are several different types of mass spectrometers. The three types most widely used today are magnetic, quadrupole, and ion trap. All of the discussion dealing with general gas chromatography (i.e., injection port liners, columns, etc.) applies to GC/MS as well. Probably the most frequently encountered problem in GC/MS is poor or inadequate sensitivity. The instrument is basically working but the response of the system for a given amount of analyte is less than expected. The problem can be anywhere from the injection port of the GC to the MS data system. Here are a list of topics with a brief explanation of how to isolate whether they are contributing to the problem.

Low Response vs. the Internal Standard

You can increase the response for low molecular weight volatile compounds by increasing the flow rate.

Suppose there is reasonable sensitivity for the internal standard; however, the response of the target compound compared with the internal standard is less than you expect. When this happens you are probably not getting the proper amount of the target analyte into the ion source. This may be caused by nonoptimal flow through enrichment devices (i.e., jet separators) since separators do not treat all compounds equally. This is especially true when analyzing for volatile analytes. The lighter the component (i.e., the lower the molecular weight) the faster it diffuses and, depending on the flow rate through the separator, you will lose proportionally more of those components. For example, you can think of two compounds, one with a molecular weight twice that of the other. If the flow rate through the separator is sufficiently slow, a higher proportion of the lower-molecular-weight-compound will be lost due to its higher diffusion rate. You will observe a low response relative to the higher-molecular-weight compound. If you increase the flow rate of the column and subsequently the flow through the separator, less of the low-molecular-weight compound will be lost since it has less time to diffuse. This will result in a higher response relative to the higher-molecular-weight compound. Therefore, you can change the relative response of low-molecular-weight compounds by adjusting the flow rate.

This is only a problem if you need to establish minimum responses like those required by the Contract Laboratory Program (CLP). If you have problems in meeting minimum response the problem may be related to the flow rate. Higher flows may help achieve required minimum responses. There is an upper limit to how high a flow rate you may use. Remember, you are using the jet separator as an enrichment device. The higher the flow the lower the enrichment. So you must balance the flow rate to get optimal response for target analytes while still maintaining good enrichment. Flow rates to the separator should be in the range of 20 to 40 ml/min. The efficiency of the separator changes with flow. Therefore, relative response will also change with flow. It is not difficult to optimize flow rates with modern mass spectrometers since they can handle relatively high flow rates due to their pumping systems. Keep in mind, however, that if you do adjust the flow rate, you will likely have to recalibrate for *all* target analytes since all the relative responses will change.

A low response condition for semivolatile compounds might indicate that the problem is in the injector. The injection port liner might be selectively absorbing the compound of interest or the splitless time might not be long enough. These issues are discussed in the Gas Chromatography section of this chapter.

Low Response

There are three possible causes of low response. Either you are not getting your sample into the ion source of the mass spectrometer, the ion source is not producing sufficient ions, or you are not detecting the ions you produce. The first cause is 10 times more likely than the second and the second is 10 times more likely than the third. Keep that in mind when you face this situation.

The Sample is not Getting to the Source

Pressure below 10^{-6} torr can indicate separator blockage

When using a jet separator it is possible to plug the separator with column packing material or other dirt that gets into the GC oven. This normally happens when changing columns. Under normal conditions the pressure in the mass spectrometer should rise to about 10^{-5} torr when the column flow through the separator is unrestricted. This is about the same pressure you will see with a capillary column flowing at about 1 ml/min. If the pressure is below about 10^{-6} torr you either have a plugged separator or a very low flow through the column. If the jet is plugged you have to cool the interface down, shut off the MS, vent the system, and remove the separator. The separator can be unplugged with gas pressure or vacuum by applying pressure opposite to the way the plug entered.

Check for jet separator reversal

Separators normally have a large and small jet. The column, either a packed or wide bore capillary, is connected on the small jet side of the separator. The bigger jet side is connected to the mass spectrometer. If you put the separator in the wrong way you will lose sensitivity. If you have low sensitivity and you are using a jet separator, make sure the separator is installed in the proper configuration. Separators normally have a mark on them indicating which side is which.

If material is getting through the injector and into the column you may have a problem with the placement of the column in the ion source of the mass spectrometer. One time we installed a capillary column in a mass spectrometer and positioned the exit of the column right at the entrance of the ion source. We then pumped the mass spectrometer down and were only able to detect the solvent when we made an injection. What happened was the column slipped during pump down such that the exit of the column was on the other side of the ion source. The column had actually moved all the way through the ion source and out the other side. The target analytes never got into the source because they went directly out the other side and were not detected.

Use whiteout to position the column

The moral of the story is to measure the distance from the end of the swage-lock nut that seals the column to the entrance edge of the ion source. The best way to make sure the column is at the right length into the source is to mark the column with whiteout. When the system is down for repair, thread the column in so it is just barely inside the source and screw the nut and ferrule holding the column in so it is finger tight. Mark the column with a little of the whiteout just below the nut. **Note:** *Some GC/MS systems may not provide an easy access to view the ion source while inserting the GC column. In this case you must rely on the manufacturer to provide the measurement.*

Remove the column and measure from the end of the column to the whiteout mark. *Write the measurement down* where you won't lose it, preferably in your notebook or instrument log book. When you reinstall the column make sure the whiteout is showing below the nut.

Leaks

Leaks in or around any fittings can cause loss of sensitivity due to high background and sample loss. If there is a leak the air and water background will be high. You know you have a leak when there are peaks at masses 28 (nitrogen) and 32 (oxygen) and they are in a ratio of 3 to 1 *and* the peak at 28 is larger than the peak at 18 (water). There is almost always a little air background in any system. The helium you buy also has some air in it, so the air background may be high when you have flow from the column going into the mass spectrometer.

Leaks can be found by spraying argon gas on suspect fittings

A good way to check for leaks is to use argon gas. Set up the mass spectrometer to scan over a few amu centered on mass 40 of argon. Spray the argon over any area where you suspect the leak. Normally fittings and connection points into the high vacuum system are where the leaks will occur. If you see a rise in the argon peak (mass 40) you will know where the leak is. Some people use a q-tip applicator soaked with methanol to find leaks. In this case you just swab the area you want to check and monitor masses between 30 to 35. An increase in this area will indicate when you find the leak.

Not Forming Sufficient Ions in the Ion Source

There are three basic components to a mass spectrometer. There is an area where you make ions — called the ion source, an area where the masses are separated or filtered — called a mass analyzer, and an area where you detect ions — called the detector or ion multiplier. If you have low sensitivity and you believe the sample is getting into the mass spectrometer then it is likely you either have a column or transfer-line alignment problem or a source problem. Sometimes, when using capillary columns that are threaded into the ion source, if the column is pushed too far into the source you may not see very much of the compounds that elute. This was discussed above. If you have an internal transfer line inside the mass spectrometer you should check it to make sure it is aligned correctly.

If the problem is really in the ion source, you can check it by running a probe sample if your instrument has a probe or perhaps by looking at the intensity of PFTBA (FC-43) spectra. If PFTBA intensity is down or the response from a known amount of a sample introduced on the probe is low, the problem is usually a dirty ion source. The first, and easiest, thing to do is to check the tuning. Try to increase the intensity of PFTBA without distorting the peaks too much. Also check the emission and collector currents if possible. The collector current should be about 50% or more of the emission current. One good approach is to keep records to establish a baseline for tuning voltages. As you notice these voltages changing you should consider source cleaning. As a last resort, shut the system down, remove the source, clean it, and put in a new filament. A new filament is a good idea because sometimes the old filament sags and is not in good alignment with the electron aperture into the ion source. If the filament sags you will notice that the collector current is low. Figure 3-13 shows a diagram of an ion source for reference.

The way you tune the ion source will affect peak shape to some extent. The voltage used to push or pull the ions into the analyzer will have a derogatory effect on peak shape. The higher the voltage the more this effect.

Analyzer

You can avoid most analyzer problems through careful cleaning procedures

There are few things that can go wrong with the analyzer. Once in operation it tends to stay that way. The rods do get dirty and need to be cleaned occasionally, but usually not as often as the source. Analyzer problems are normally caused as a result

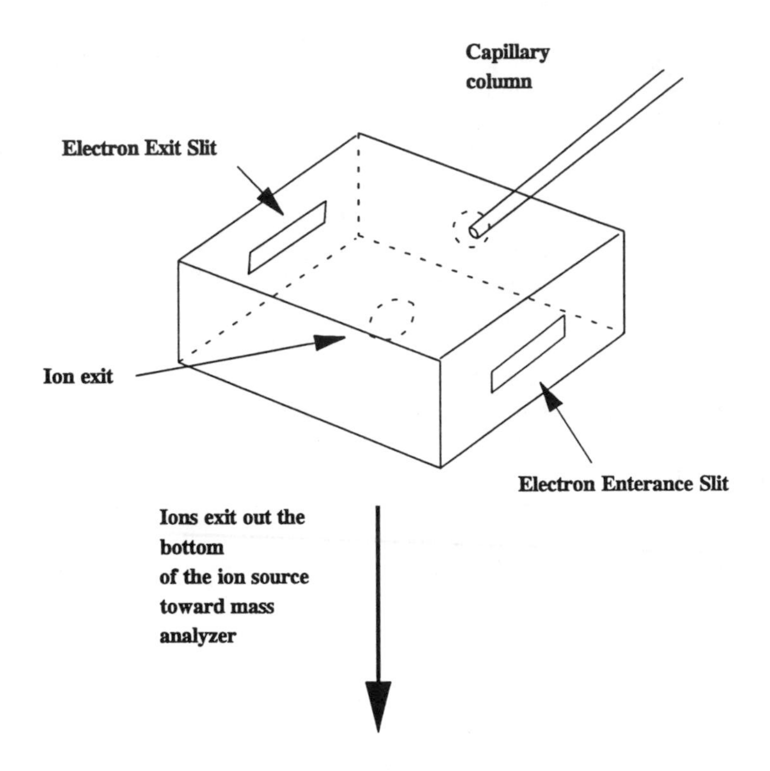

FIGURE 3-13. Diagram of an ion source electron impact.

of improper cleaning or reassembly after cleaning. A dirty analyzer (rods) will cause loss of mass resolution.

Resolution for quadrupole instruments is controlled by the ratio of the radio frequency (Rf) and direct current (DC) voltages applied to the rods. If we need a ratio of 10:1 Rf to DC to maintain unit mass resolution then for every 100 V Rf we need 10 V DC. If this ratio changes you will see a change in the mass resolution. Figure 3-14 shows an example of a spectrum when the DC voltage is not high enough for the Rf to maintain unit resolution. In this case you get lots of ions coming through the quadrupole filter; however, they are unresolved. This may result from a short circuit in the DC supply or one of the rods might be arcing. In any event something is wrong with the Rf to DC ratio.

Another thing that can go wrong in the analyzer is incorrect mass assignment or mass assignment drift. This is normally caused by large temperature fluctuations in the room where the mass spectrometer is. The temperature affects how the Rf is generated and may cause misassignment.

About the only other thing that can go wrong with the analyzer is putting it back together the wrong way. That is, there is a front, which goes toward the source, and a back, which goes toward the detector. If you get them reversed you won't achieve the same resolution you were seeing before. The peak shapes will

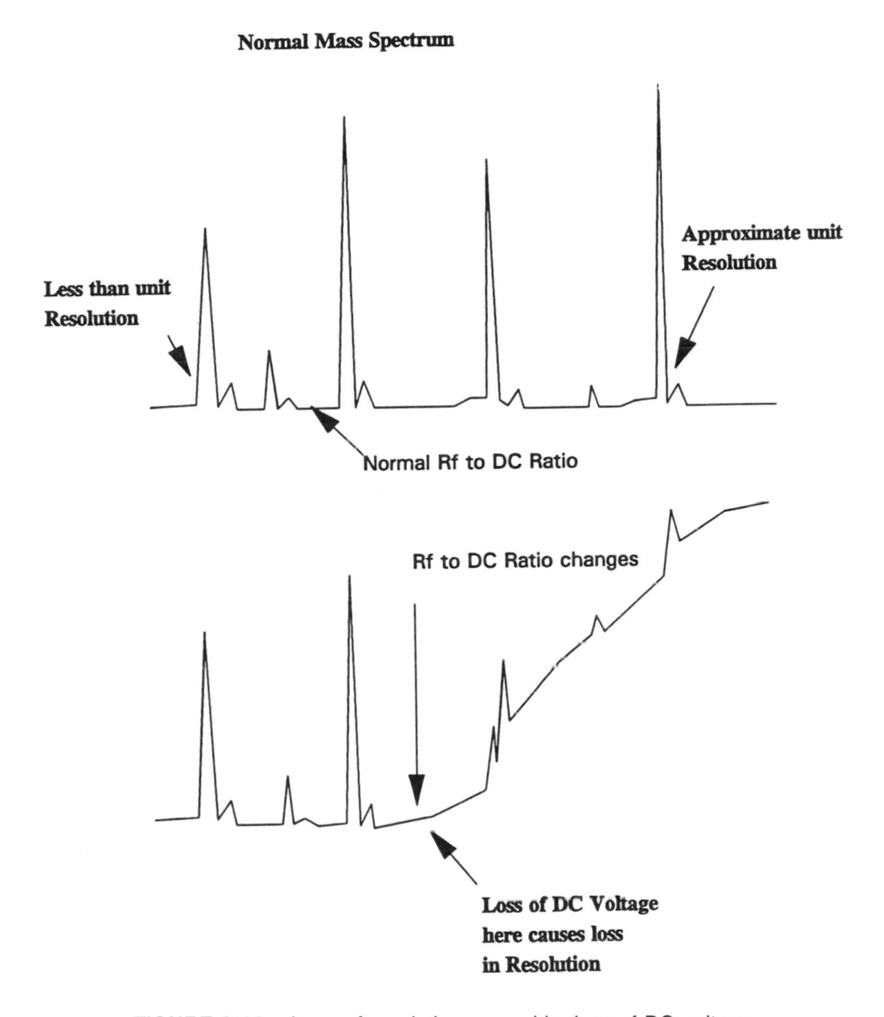

FIGURE 3-14. Loss of resolution caused by loss of DC voltage.

be different, etc. Be sure you know which end is which whenever you remove the rods for cleaning.

Detector

Like the analyzer, there is very little that can go wrong with the detector. The gain of the detector will go down with age and use. Typical gains are in the range of 10^5. The drop off in gain is proportional to usage. That is, the higher the cumulative current through the multiplier the more likely it will need to be replaced. You can think of a new multiplier like an empty electron jar. As you use the multiplier you add more electrons to the jar. The higher the voltage on the multiplier the faster you fill the jar up. The jar will only hold a given amount of electrons. As you reach that limit you will notice the gain falling off. You will have to increase the voltage to get the same multiplication. Eventually, the gain will fall to 10^4 or 10^3 and the multiplier should be replaced.

Assuming there is voltage to the multiplier you will see some mass peaks. They may be small, but indicate that the multiplier is working. If you replace a multiplier, the alignment should be checked to insure the ions will be attracted and collected.

Scan Speed

The rate of change of compound into the source should be comparable to scan speed

Scan speed is related to several problems that you will likely observe at some time. Specifically, sensitivity and spectral quality are directly related to scan speed. This is especially important when using capillary columns because the chromatic peak width is much narrower than with packed columns. This means that, for a given scan time, you will have fewer scans across the peak. Fewer scans mean the possibility of skewed spectra. Figure 3-15 gives some examples of this. If you scan too slowly relative to the width of the peak you will lose chromatographic resolution and generate nonrepresentative spectra. This is generally not a problem for packed columns since the chromatographic peaks are fairly wide — usually 10 sec or more. This, however, is not the case for the capillary columns.

If we are scanning at a rate of 1 sec/scan over a range of about 500 amu and the compound has ions both at the high and low ends of the mass range, then the spectra at the peak maximum may be skewed. This happens as follows: the first two scans are skewed because the higher mass ions are greater in relative intensity since the amount of compound entering the ion source is increasing. This means that ions at higher masses will be proportionally greater than they would be if there were a constant amount of the compound in the source. The scan at the top will also be skewed because the concentration of the compound is rising until the middle of the spectrum and then the concentration in the source starts to fall. This means that the higher masses will be smaller, relative to the lower masses, than they should be.

The rule of thumb that most use is to try to get about 10 scans across a gas chromatographic peak. To do this you estimate the peak width and then divide by 10. If the peak width is 5 sec then you should scan at a rate of about 0.5 sec per scan. You can increase your scan speed but if you scan too fast you lose sensitivity since you spend less time on each mass. If you scan at 0.25 sec per scan instead of 1 sec your sensitivity will fall off by a factor of 1.4, as shown in the following equation.

$$\Delta\text{Sensitivity} = \sqrt{\frac{\text{scanspeed 2}}{\text{scanspeed 1}}} \quad (1)$$

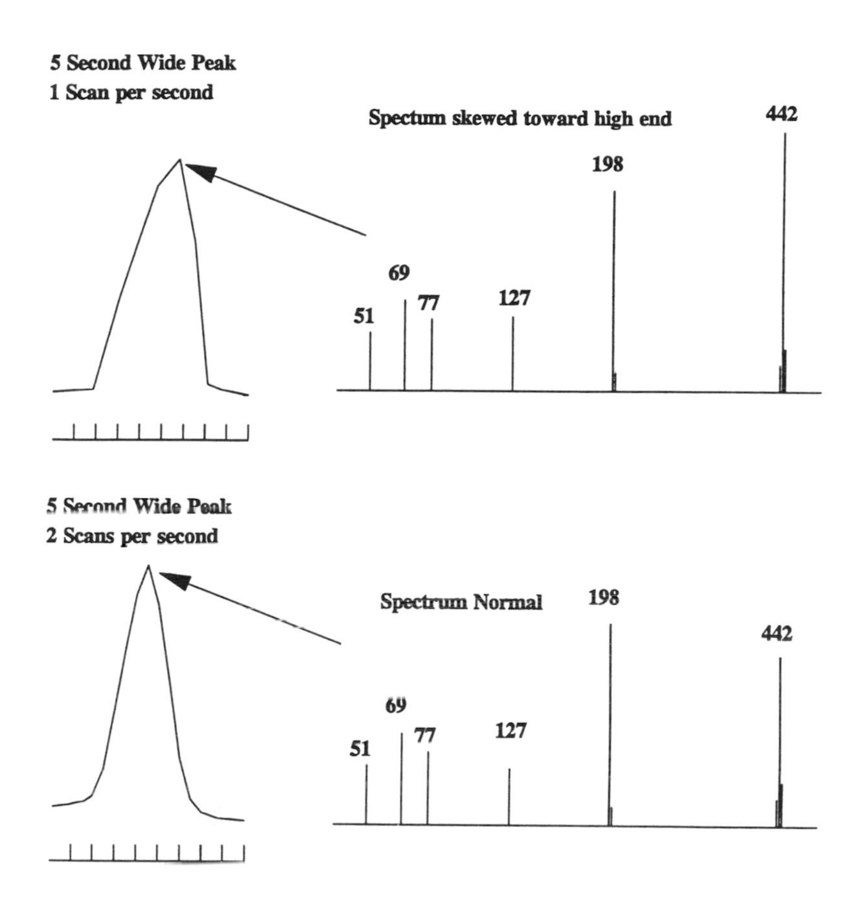

FIGURE 3-15. The effect of scan speed on DFTPP.

This equation also shows how multiple ion detection increases sensitivity. You spend longer on a few ions and your signal to noise ratio (i.e., the increase in sensitivity) goes up for those ions. Using the same example, if you scan at 500 amu/sec you spend about 2 msec on each mass. If you only scan 10 ions in 1 sec (i.e., 100 msec for each mass) then your sensitivity will go up by about a factor of 7:

$$7 \approx \sqrt{\frac{100}{2}} \tag{2}$$

More ADC bits do not provide a better dynamic range

Another issue related to scan speed and sensitivity is dynamic range. Every mass spectrometer must convert the analog output from the electron multiplier into a digital signal for the data system. This conversion is accomplished by an analog to digital converter — or simply the ADC. The more bits the ADC has the better the *Intensity Resolution,* not the dynamic range. All converters used in mass spectrometry rely on a 0- to 10-v scale. If you use a 12-bit ADC you divide the range into 4096 segments. If you have a 16-bit ADC you divide the range into 65536 segments, or 16 times more resolution, *not dynamic range.*

Electron Energy

Spectra are also affected by the ionization energy used. Typically, electron impact ionization is performed at 70 eV. In fact, the electrons that impact the sample molecules are in a range of energy centered at the selected voltage. That is, the electron energy is distributed about 70 eV. Some electrons are more energetic and some are less energetic. As long as the distribution is not too large this isn't a problem. When the electron interacts with the molecule during ionization it only transfers a portion of its energy to the molecule. The resulting molecules (actually now they are ions) will also have an energy distribution. Some molecules fragment to give the characteristic mass spectrum while others are detected intact — these give rise to the parent peak in the mass spectrum. The 70-eV standard came about simply because it is the *lowest* voltage that produces the most ion current. If you can adjust the ionizing voltage you will not see any signal below about 12 eV. As you increase the voltage the ion current will steadily rise until you get to about 70 eV. Above 70 eV you will not see much of an increase in ion current. The spectra obtained as you increase the voltage, however, will change dramatically.

You can detect "missing" molecular ions by lowering the ionization energy

The lower the ionization energy the less fragmentation you will get. This is a good technique to use to get a better idea of the molecular weight of a compound. If you suspect that you are not seeing the molecular ion for a given compound, you can lower the ionization energy and see if the molecular ion can be established. Electrons accelerated by lower voltages have less energy. Thus, the electron has less energy to transfer to the compound of interest. This, in turn, means there will be less fragmentation and, therefore, a better chance of observing the molecular ion. The higher masses will come up relative to the lower masses. The overall ion current will go down as a result and your sensitivity will also go down. However, this is a very useful technique for establishing the molecular ion despite the loss in sensitivity.

Ion Source Temperature

The temperature of the ion source dramatically affects the spectra you obtain. For many compounds, the lower the source temperature the fewer mass fragments you will see. This is especially true for compounds that are of low to moderate stability. For example, Decaflurotriphenylphospine (DFTPP) fragmentation is susceptible to source temperature effects and, therefore, if the source temperature is not high enough the spectra you obtain will not meet tuning specification.

For most electron impact mass spectrometers a source temperature of about 170 to 200°C is optimal. Some mass spectrometers do not display source temperature directly. In this case you have to experiment with instrument parameters to obtain the optimal amount of fragmentation. It is usually a safe bet that if mass 442 is the base peak of DFTPP the source temperature is not high enough and should be increased. Put another way, the amount of energy imparted to the DFTPP molecule is not high enough to induce the required amount of fragmentation and one way to get more energy into the molecule is by increasing the temperature.

Instrument Zero

Noise volume is inversely proportional to zero setting

If the instrument zero adjustment is set too high or too low it can cause several problems. If the zero is set too low the resulting spectra will be noisy and, more important, you will use up your data system disk space very quickly. If you see a mass peak at every amu the zero is probably set too low. You can expect to see more mass peaks at the lower weights; however, you should not see masses at every amu. In the mass range above 100 amu you should only see an occasional peak. If you are running in the multiple ion detection (MID) mode, however, you should adjust the instrument zero to make sure you see some signal from each scan, since this will establish the noise level. (You will need this to ensure correct quantification and to establish your signal to noise ratio.)

If you have the zero set too high you will lose some of the lower intensity ions and ion ratios will not be correct. For example, if a compound has 10 carbon atoms it will have an M + 1 ion at about 11% of the molecular ion. If the zero is set at 1% the M + 1 will be 10%. One could confuse a 10% value for 9 carbon atoms instead of the correct value of 10. While this is not generally a problem for environmental labs, you should check isotope ratios from time to time to make sure you are getting reasonable values.

You can usually tell when the zero is set too high by looking at the resulting chromatogram. If there is little or no noise in the chromatogram then the zero is probably too high. Look at Figure 3-16. What can you say about the zero setting?

Tuning

There have been several methods developed over the years to tune a mass spectrometer to meet DFTPP tuning criteria. The following is a brief description of the method used by Aznavoorian et al. for tuning instruments in the Finnigan 4000 family; however, it should work for other types of mass spectrometers as well.

You introduce PFTBA into the ion source and use the ions at 112 or 114 and the ion at 414 to tune the instrument.

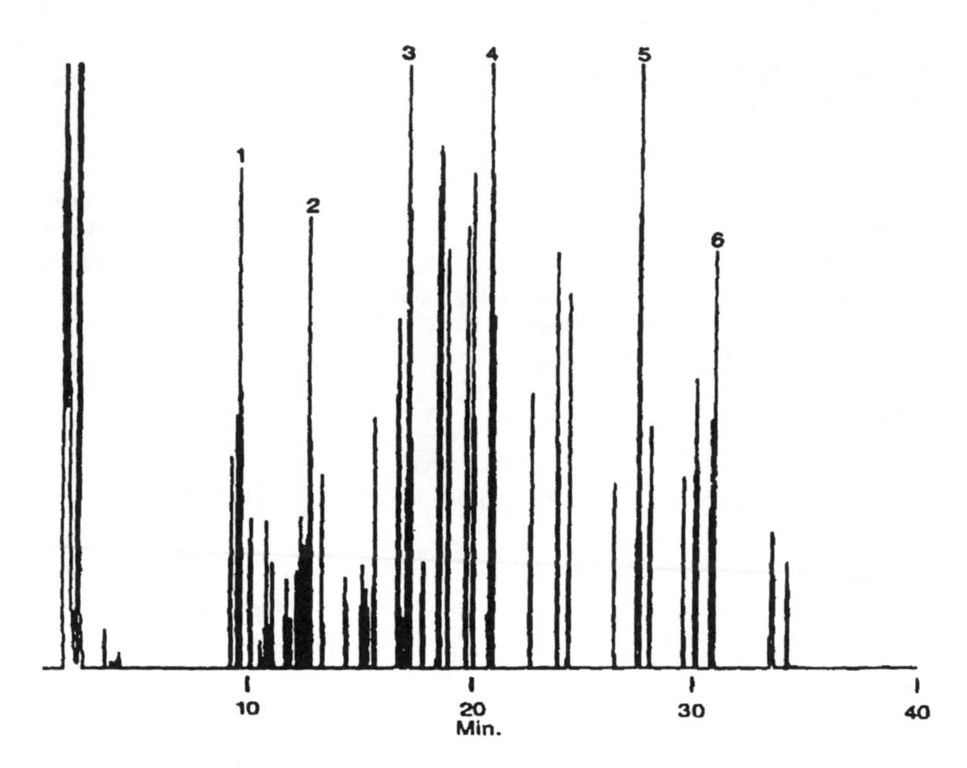

FIGURE 3-16. Reconstructed ion chromatogram. Instrument zero is set too high.

Note: *Use the lowest pressure possible for the PFTBA.* When you open the PFTBA valve to introduce the compound the pressure in the instrument may go up. This increase in pressure will affect how the instrument tunes. In other words, optimal lens and source voltages are a function of pressure in the ion source. Therefore, use the smallest amount of PFTBA possible while tuning the instrument. This will ensure optimal voltages under GC/MS conditions.

First, you tune the instrument so the ratio of 114 (or 112) to 414 is about 2 to 1. At this point you inject DFTPP to see if the criteria are met. If not, continue to increase the ratio up to about 4 to 1. Obviously, if the ion at 442 is less than 40% you need to decrease the 114/414 ratio. If the ion at 51 is less than 30% you need to increase the ratio. If both 51 and 442 are below the criteria you need to reduce the ion intensity in the middle of the mass range to get 198 ion intensity down. You can do this by looking at the PFTBA ions at 70, 220, and 414. Try to reduce the height of 220 while maintaining the heights of 70 and 414. If you can reduce the 220 by 20% relative to 70 and 414 you should be in the proper range for DFTPP.

Once you meet DFTPP you should write down the intensities of 69, 100, 114, 131, 169, 219, 264, and 414 ions of PFTBA or simply note the ratios of 114/414, 131/69, and 219/69. If the operator can come close to these ratios then meeting DFTPP should not be a problem. By using this procedure the initial calibrations can last as long as

14 weeks. Operators with minimal experience can successfully tune the instrument within about 20 min.

Data Reduction

When processing data for target analytes (which is typical for environmental analyses) the data processing occurs in two distinct stages. Both occur at the instrument data system and they both ultimately influence the accuracy of the determination. The first part of the processing involves deciding whether a target analyte is present. Compounds that are present are then quantitated. Errors can be made in either or both steps. For example, a compound may be present but is not identified. This type of error is called a false negative. Conversely, a compound may be identified as present when in fact it is not. This type error is called a false positive. These errors are compounded through the quantitation process.

Compound Identification — False Positives and Negatives

You might ask how false positives and negatives occur and how you might avoid them. Before discussing this we need to understand how the GC/MS data system actually performs its task of identifying compounds. Once this has been accomplished we can turn our attention on how to avoid either of these conditions.

Target compounds are identified by the GC/MS from a target compound library which the GC/MS operator makes up. The way this is done is the operator analyzes one or more standards that contain the target compounds at a concentration that makes them easy to identify. Typically, 50 ng of each component delivered onto the GC column will provide enough signal to easily identify most compounds. The operator identifies each component based on its expected retention behavior and resulting mass spectra. For example, when establishing a library for a volatile analysis the operator knows that *cis*-1,3-dichloropropene elutes first, followed by toluene, *trans*-1,3-dichloropropene, 1-chloro-2-bromopropane, 1,1,2-trichloroethane, etc. Furthermore, the operator knows that the spectra for toluene will contain a large peak at 91 and 92 while the spectra for dichloropropene will contain a peak at 61. Using this type of scheme the operator can quickly establish the identity of each GC peak and obtain good spectra for each compound. Spectra at the top of the peak are background subtracted and then reduced for insertion into the target compound library.

A 10:1 ratio between ion counts and background ions is sufficient for a library entry

Background subtraction may be performed automatically (i.e., by the data system) or manually. Either approach provides acceptable results assuming the original spectra are intense. This means that the ion intensities of each mass in the spectra are sufficient to withstand background subtraction without degrading the spectra of the target compound. For example, if we have some background ions that are about

100 ion counts and the target compound only produces ions of 500 ion counts, then the subtracted spectra may not reasonably represent the target compound. You definitely want to avoid this condition when you are building a target compound library. If the ion counts for the target compound are 10 times (or more) greater than the background ions, the background subtracted spectra should be of a sufficient quality (intensity) for insertion into a target compound library.

After obtaining a background subtracted spectra for the target compound the data system and/or the operator "reduces" the spectrum for insertion into the library. Data reduction is necessary for reasons of speed and to eliminate false negatives — as explained in the following example.

A critical step — the number of ions stored

Let's say a target compound produces spectrum that has 50 ions and that these ions vary in relative intensity from 0.5 to 100%. Obviously the ions of highest intensity will be more useful in compound identification and the most intense ions should become part of the library entry. Furthermore, many GC/MS data systems can only store a limited number of these ions. Let's say you direct the data system to keep only 3 ions of the original 50. This means that the possibility of a false positive is higher than if you stored more ions — say 10. Some data systems will allow you to store only 5 ions while others will allow up to 50. The more ions you store the higher the level of confidence in subsequent target compound identification. Unfortunately, the more ions you store the slower the data system runs and the greater the possibility of a false negative.

A false negative can occur if you direct the data system to store too many ions or make the identification criteria too restrictive. At first you might think that keeping more ions in your target compound library will improve the identification of the target compounds. This is true up to a point. Let's say you decide to keep 25 of the ions from the original spectrum. When you try to match that spectrum to an unknown you will have to get a fairly close fit with all 25 of the ions before you will call the compound present. The more ions you include the higher the likelihood that some of them will not match and, therefore, of not calling the compound present — even when it is! The unknown spectrum never matches the library spectrum exactly. This is due to interferences present in the unknown spectrum and differences in mass spectrometer tune. From the time the library was created you may have had to adjust the mass spectrometer, cleaned it, etc. This will cause small changes in the resulting spectra compared with the library. So you see, there is a trade off between the number of ions stored and the possibility of false positives, false negatives, and data system speed.

Generally, you will find that most systems perform adequately by storing between 3 and 12 ions in the target compound library. If you store more ions the data system will run more slowly and you run a greater risk of getting false negatives. If you only store one or two ions you will get more false positives because it is likely that some compounds in the sample will have the same ions as one or more target compounds. Each lab has to decide how best to handle this situation. However, most lab managers are probably unaware of this critical decision and leave it up to their operators to set the libraries and identification criteria.

The library spectrum should be updated periodically to account for any changes in the instrument. The operator updates the response factors and retention times on a daily (or more frequent) basis. The library spectrum, however, is almost never updated unless the library is lost (i.e., the library is accidentally erased, major hardware failure, etc.). The library spectra should be revised periodically as part of the laboratory's QA program and as part of establishing the false positive and negative rate for the GC/MS lab.

Quantification and True Detection Limits

It is always amusing to watch someone use detection limit information. The following scenario is fairly typical. The laboratory analyzes samples for a client and reports results. Many of the target compounds are not found and the laboratory reports that the concentration for these compounds is below the detection limit for each compound. The detection limit or, more accurately, what the laboratory believes is the detection limit, is reported for each compound. The client then takes the detection limit values and plugs them into a risk estimation model and finds that the potential risk of exposure to the chemical is above (or below) some theoretical limit. If the value is below the limit then the client breaths a sigh of relief and goes on evaluating the data. If the value is above the threshold the client will call the lab and ask for lower detection limits. This situation is totally absurd and would actually be funny if it were not so common — and expensive. The point is that the detection limits reported by the lab in the first place are wrong. Remember the detection limit rule? In fact, the lab has no idea what the actual detection limit is for the client's samples.

Pitfalls of detection limits

Some labs report detection limits based on "contract required" detection limits established by EPA's Contract Laboratory Program. These limits are entirely arbitrary and represent concentrations that the lab "should" be able to detect in a clean (water) sample. Whether the lab can actually detect this concentration in an actual sample is unknown.

The only way the lab can establish a detection limit for a given sample is to spike that sample at several concentrations and identify what it can, in fact, detect in that sample. This situation is related to the false-negative condition we discussed above and is particular to GC/MS data. Since quantitation is not performed unless the compound is identified as being present the entire matter of detection limit is, in reality, undefined.

The point of this discussion is the following. Don't believe detection limits and especially, don't try to use detection limit data to estimate risk unless the lab actually determined the detection limit in the samples by spiking them. Labs should make this clear to their clients.

Avoiding Reanalysis

One of the most time-consuming operations in a GC/MS lab is reanalysis. Samples may be too concentrated and contaminate the GC system, and possibly the mass spectrometer. If this happens, you may need to shut the system down and clean it — losing 4 to 8 h in the process. Often a sample will contain one or two components at a high (ppm) concentration while the rest of the components are below detection limits. The problem is to keep everything on scale in one run. Let's examine what you can do about this problem.

As indicated above, it is fairly common to have one component in a sample at a much higher concentration than other components. Since the mass spectrometer is typically calibrated only over one order of magnitude (e.g., from 20 to 200 ng) samples that contain more than 200 ng of a component must be diluted if you want the result to be in the calibrated range. Of course you can calibrate the mass spectrometer over a wider range. Suppose we calibrate up to 2 µg. Now we are calibrated over two orders of magnitude. You could possibly go a little higher; however, if you do two things will happen. You will overload the GC column and you will saturate the detector. Overloading the column is not a major problem, but your chromatography will suffer. You can always tell when you overload a column from the shape of the peak (Figure 3-8). You can still quantitate mass peaks if you have a column overload. It just doesn't look very good.

Calibrate over two orders of magnitude instead of one

Overloading the column may cause a shift in retention behavior for compounds eluting later than the compound that overloaded the column. What happens in this case is as follows. The compound that overloads the column is at such a high concentration that it changes the behavior of the liquid phase in the column. This in turn changes how the later compounds elute. The compounds that elute before the overload do not experience any problems since they are ahead of the overload. The liquid phase in the column acts normally. As the overloading compound comes by it dissolves in the liquid phase, the liquid phase becomes saturated with the compound and you no longer have an equilibrium between the gas and liquid phases. This means that the liquid phase is really a mixed phase consisting of the original liquid phase plus the overloading compound. Compounds that come after the overload experience a different liquid phase and hence their retention behavior will be different. Again, this is not a terrible problem to overcome, but you must understand that you will need to compensate for the shift in retention behavior for the later eluting compounds.

How to reduce reanalysis when the compound is over-concentrated

More often you will have a situation where one compound in the mixture will be outside of the calibration range or the principal quantitation ion will be saturated. Saturation means that the compound is so concentrated that it exceeds the range of the detector. Most labs just dilute the sample and reanalyze it. This works just fine but you lose time and money with every reanalysis. There is a way to avoid reanalysis when this happens. The way to do this is to use multiple library entries for each compound. Here is how it works.

Create multiple library entries of different ions for "suspect" compounds

Let's say you obtain a spectrum of toluene. The molecular weight of toluene is 92; however, the base peak (i.e., the peak that is the most abundant in the spectrum) is 91. This base peak has the formula of C_7H_7 and is the ion that is normally used to quantitate toluene. Let's suppose that the concentration of toluene in a sample is high and the ion at mass 91 goes off scale (i.e., saturated). In this situation most labs would reanalyze a smaller sample. This takes double the amount of time. Another way to get toluene to be "on scale" is to use a different ion to quantitate. Toluene has ions at 92 and 77 that are lower in abundance than the base peak. Let's suppose that the peak at mass 77 is only 10% of the mass at 91. If we create another library entry for toluene and designate the quantitation ion as 77 instead of 91 we can quantitate toluene even if the ion at 91 goes off scale. This means that from one analysis you can achieve a valid quantitation even when a compound is outside of the normal concentration range. It only requires that you create an additional entry in the library for compounds that frequently go off scale. Since toluene is a fairly common solvent you might expect that a significant number of samples will contain toluene in a high enough concentration that requantitation is required.

Creating additional entries in the library is simple. All you do is make a copy of the original entry and change the quantitation ion to one that is 10% (or less) of the base peak. This assumes that the base peak is used as the normal quantitation ion. In any case you should look at the spectrum and select an ion that is approximately 10% or less of the quantitation ion. Add this new entry to the library and update the response factor since it will be higher than the normal quantiation ion. Once this is done all you have to do is make sure you keep this entry current by updating the retention time and response factor each day when you run the calibration standard. That's it. By employing additional library entries you should be able to eliminate about half of the reruns you are making.

APPENDIX 1

Example ASCII Files from GC/MS Data Systems for Export to LIMS

V2018.MSQ

11/23/92
Quantitation Report File: 2018

Data. 2018.TI
10/24/84 12:04:00
Sample: CLP, 3134, 3134-0001, SAMP01, L, W, DEMO_SAMP, VOA, EPA,
Conds.: PACK, 2016, VVV, 2015, 2019, 2020
Submitted by: FINNIGAN Analyst: WCS

AMOUNT = AREA * REF.AMNT/(REF.AREA)* RESP.FACT)
Resp. fac. from Library Entry

NO NAME

1	CI01	BROMOCHLOROMETHANE*****INTERNAL STANDARD******
2	C010	CHLOROMETHANE
3	C015	BROMOMETHANE
4	C020	VINYL CHLORIDE
5	C025	CHLOROETHANE
6	C030	METHYLENE CHLORIDE
7	C035	ACETONE
8	C040	CARBON DISULFIDE
9	C045	1,1,-DICHLOROETHENE
10	C050	1,1-DICHLOROETHANE
11	C055	TRANS-1,2-DICHLOROETHENE
12	C060	CHLOROFORM
13	C065	1,2-DICHLOROETHANE
14	CS15	1,4-DICHLOROETHANE-D4****SURROGATE****
15	CI10	1,4-DIFLUOROBENZENE****INTERNAL STANDARD#2****
16	C115	1,1,1-TRICHLOROETHANE
17	C110	2-BUTANONE
18	C125	VINYL ACETATE
19	C120	CARBON TETRACHLORIDE
20	C130	BROMODICHLOROMETHANE
21	C140	1,2-DICHLOROPROPANE

NO	NAME	
22	C145	TRANS-1,3-DICHLOROPROPENE
23	C150	TRICHLOROETHENE
24	C165	BENZENE
25	C155	DIBROMOCHLOROMETHANE
26	C160	1,1,2,-TRICHLOROETHANE
27	C170	CIS-1,3-DICHLOROPROPENE
28	C175	2-CHLOROETHYLVINYLETHER
29	C180	BROMOFORM
30	CS05	TOLUENE-D8****SURROGATE****
31	CI20	CHLOROBENZENE D5 ****INTERNAL STANDARD****
32	C215	4-METHYL-2-PENTANONE
33	C210	2-HEXANONE
34	C220	TETRACHLOROETHENE
35	C225	1,1,2,2-TETRACHLOROETHANE
36	C230	TOLUENE
37	C235	CHLOROBENZENE
38	C240	ETHYLBENZENE
39	C245	STYRENE
40	C250	TOTAL XYLENES
41	CS10	BROMOFLUOROBENZENE****SURROGATE****

No	m/z	Scan	Time	Ref	RRT	Meth	Area(Hght)	Amount	%Tot
1	128	221	9:24	1	1.000	A BB	133824.	50.000 UG/L*	6.87
2	NOT	FOUND							
3	NOT	FOUND							
4	NOT	FOUND							
5	NOT	FOUND							
6	49	164	6:58	1	0.742	A BV	21138.	2.521 UG/L	0.35
7	43	175	7:26	1	0.792	A BV	716600.	413.641 UG/L	56.80
8	NOT	FOUND							
9	NOT	FOUND							
10	63	235	9:59	1	1.063	A BB	243.	0.030 UG/L	0.00
11	96	247	10:30	1	1.118	A BB	287.	0.079 UG/L	0.01
12	83	259	11:00	1	1.172	A BB	611.	0.085 UG/L	0.01
13	NOT	FOUND							
14	65	270	11:28	1	1.222	A BV	217831.	52.216 UG/L	7.17
15	114	405	17:13	15	1.000	A BB	594496.	50.000 UG/L*	6.87
16	97	296	12:35	15	0.731	A BB	207.	0.059 UG/L	0.01
17	NOT	FOUND							
18	NOT	FOUND							
19	117	304	12:55	15	0.751	A BB	323.	0.092 UG/L	0.01
20	NOT	FOUND							
21	NOT	FOUND							
22	NOT	FOUND							
23	NOT	FOUND							

No	m/z	Scan	Time	Ref	RRT	Meth	Area(Hght)	Amount	%Tot
24	78	359	15:15	15	0.886	A BB	24698.	1.704 UG/L	0.23
25	127	364	15:28	15	0.899	A BB	217.	0.050 UG/L	0.01
26	97	364	15:28	15	0.899	A BB	299.	0.060 UG/L	0.01
27	NOT	FOUND							
28	NOT	FOUND							
29	NOT	FOUND							
30	98	473	20:06	15	1.168	A BV	543908.	51.286 UG/L	7.04
31	117	498	21:10	31	1.000	A BB	455477.	50.000 UG/L*	6.87
32	NOT	FOUND							
33	NOT	FOUND							
34	NOT	FOUND							
35	NOT	FOUND							
36	NOT	FOUND							
37	NOT	FOUND							
38	NOT	FOUND							
39	NOT	FOUND							
40	NOT	FOUND							
41	95	611	25:58	31	1.227	A BB	623316.	56.391 UG/L	7.74

RVXNSP.RR

```
11/23/92
 .QN   VXNSP                                                            0012A07
                 96010311861610R   36   240   2.325  9                  0020A4A
                 CI01   Bromochloromethane                              0036807
                   1  128  260  11.1307   1    35664.UG/L     50.0000   0043A9B
                 C030   Methylene Chloride                              0054FAC
                   2   84  177   7.9115   1    13411 UG/L     14.4549   00617B9
                 C035   Acetone                                         0075F45
                   3   43  190   8.4157   1   176466.UG/L    210.7556   00846B6
                 CS15   1,2-Dichloroethane d4                           0090AC6
                   4   65  332  13.9231   1    59877.UG/L     48.2638   01031C3
                 C110   2-Butanone                                      0110997
                   5   72  330  13.8456   1     2723.UG/L      9.5831   0122E90
                 CI10   1,4-Difluorobenzene                             0133699
                   6  114  528  21.5247   6   152251.UG/L     50.0000   01444AD
                 CI20   Chlorobenzene-d5                                0150555
                   7  117  660  26.6442   7    99706.UG/L     50.0000   0163ABB
                 C230   Toluene                                         0176F52
                   8   92  628  25.4031   7      737.UG/L       .3885   01801A7
                 CS05   Toluene d-8                                     0194B5F
                   9   98  625  25.2867   7   134687.UG/L     49.4985   0205ACC
                 CS10   Bromofluorobenzene                              0217E06
                  10   95  814  32.6170   7    60343.UG/L     47.5713   0221FD8
ENDFILE                                                                 0237BE0
```

```
RVX100.RR

11/23/92
 .QN   VX100                                                          0016CE4
             96011216851422R    36    256   2.326    9                002134E
             CI01     Bromochloromethane                              0036807
               1    128  219    9.2620    1      15348.UG/L    50.0000  0041EAE
             C010     Chloromethane                                   0054E21
               2     50   18    1.4633    1      41755.UG/L   100.0000  0061798
             C015     Bromomethane                                    0070126
               3     94   39    2.2781    1      33753.UG/L   100.0000  0081F9E
             C020     Vinyl Chloride                                  0093B27
               4     62   56    2.9377    1      45877.UG/L   100.0000  01026A4
             C025     Chloroethane                                    0110220
               5     64   79    3.8300    1      23288.UG/L   100.0000  0122399
             C030     Methylene Chloride                              0134FAC
               6     84  132    5.8862    1      45233.UG/L   100.0000  01424AA
             C035     Acetone                                         0155F45
               7     43  148    6.5070    1      28487.UG/L   100.0000  01623AF
             C040     Carbon Disulfide                                0175282
               8     76  172    7.4382    1     131764.UG/L   100.0000  01837AE
             C045     1,1-Dichloroethene                              0196886
               9     96  206    8.7575    1      37111.UG/L   100.0000  02026AF
             C050     1,1-Dichloroethane                              0216882
              10     63  240   10.0769    1      84661.UG/L   100.0000  02237B1
             C053     1,2-Dichloroethene(total)                       0235E27
              11     96  259   10.8142    1      45241.UG/L   100.0000  0242EBB
             C060     Chloroform                                      0253EE2
              12     83  275   11.4350    1      86950.UG/L   100.0000  02630BE
             CS15     1,2-Dichloroethane-d4                           0270AC6
              13     65  292   12.0947    1      30121.UG/L    50.0000  02828AE
             C065     1,2-Dichloroethane                              0296982
              14     62  295   12.2112    1      68516.UG/L   100.0000  03031B7
             C110     2-Butanone                                      0310997
              15     72  292   12.0947    1       8568.UG/L   100.0000  03244A2
             C115     1,1,1-Trichloroethane                           0335693
              16     97  328   13.4917    1      73736.UG/L   100.0000  0343BC4
             C120     Carbon Tetrachloride                            035671F
              17    117  339   13.9187    1      73296.UG/L   100.0000  0364CC4
             C125     Vinyl Acetate                                   03721ED
              18     86  343   14.0739    1      12119.UG/L   100.0000  0383BB4
             C130     Bromodichloromethane                            039314B
              19     83  354   14.5008    1      78567.UG/L   100.0000  0403DBF
             CI10     1,4-Difluorobenzene                             0413699
              20    114  493   19.8952   20      66133.UG/L    50.0000  04249C1
```

C140	1,2-Dichloropropane						0433398
21	63	389	15.8589	20	57082.UG/L	100.0000	04451C2
C143	cis-1,3-Dichloropropene						0454EEE
22	75	396	16.1306	20	73455.UG/L	100.0000	04649BB
C150	Trichloroethene						047276E
23	130	411	16.7128	20	57232.UG/L	100.0000	04850B4
C155	Dibromochloromethane						0492C50
24	127	428	17.3727	20	55391.UG/L	100.0000	05062B7
C160	1,1,2-Trichloroethane						0515793
25	97	431	17.4893	20	43939.UG/L	100.0000	05248CC
C165	Benzene						0535F4D
26	78	424	17.2173	20	147858.UG/L	100.0000	05459C7
C172	trans-1,3-Dichloropropene						0552145
27	75	430	17.4505	20	74585.UG/L	100.0000	0564DBB
C175	2-Chloroethylvinylether						0571489
28	63	459	18.5759	20	36885.UG/L	100.0000	05856C9
C180	Bromoform						0594197
29	173	499	20.1282	20	58077.UG/L	100.0000	06062BD
CI20	Chlorobenzene-d5						0610555
30	117	623	24.9402	30	46653.UG/L	50.0000	06244BE
C205	4-Methyl-2-Pentanone						06323AB
31	43	511	20.5939	30	69315.UG/L	100.0000	0644BB5
C210	2-Hexanone						0651387
32	43	552	22.1846	30	46003.UG/L	113.1690	06655B5
C220	Tetrachloroethene						06773B3
33	164	561	22.5340	30	53784.UG/L	100.0000	06856BC
C225	1,1,2,2-Tetrachloroethane						06935E4
34	168	561	22.5340	30	39075.UG/L	100.0000	07060B4
C230	Toluene						0716F52
35	92	595	23.8536	30	93289.UG/L	100.0000	07250CA
CS05	Toluene d-8						0734B5F
36	98	590	23.6594	30	65864.UG/L	50.0000	07444CC
C235	Chlorobenzene						0754133
37	112	626	25.0568	30	119469.UG/L	100.0000	07673BB
C240	Ethylbenzene						0770C27
38	106	679	27.1135	30	60776.UG/L	100.0000	0785DC2
C245	Styrene						0797B54
39	104	782	31.1106	30	119118.UG/L	100.0000	08073AA
C250	Xylene(total)						0812BE4
40	106	816	32.4315	30	74664.UG/L	100.0000	08255BC
CS10	Bromofluorobenzene						0837E06
41	95	745	29.6745	30	37674.UG/L	50.0000	0843DD0

ENDFILE 0857BE0

```
RVXSPK.RR

11/23/92
 .QN   VXSPK                                                           0012A04
            96010311861658R    36   232   2.325    9                   0021051
            CI01     Bromochloromethane                                0036807
              1   128  259   11.1085    1     34850.UG/L     50.0000   00432AA
            C030     Methylene Chloride                                0054FAC
              2    84  176    7.8894    1     10564.UG/L     11.6522   0061CBC
            C035     Acetone                                           0075F45
              3    43  189    8.3936    1    187672.UG/L    229.3744   00852BC
            C045     1,1-Dichloroethene                                0096886
              4    96  245   10.5655    1     39770.UG/L     53.3928   01036BD
            CS15     1,2-Dichloroethane-d4                             0110AC6
              5    65  331   13.9009    1     57816.UG/L     47.6910   01231B9
            C110     2-Butanone                                        0130997
              6    72  329   13.8233    1      3871.UG/L     13.9414   0142BA6
            CI10     1,4-Difluorobenzene                               0153699
              7   114  527   21.5025    7    147369.UG/L     50.0000   01647B3
            C150     Trichloroethene                                   017276E
              8   130  448   18.4385    7     58186.UG/L     52.5766   01847CA
            C165     Benzene                                           0195F4D
              9    78  461   18.9428    7    149235.UG/L     51.9749   02053C8
            CI20     Chlorobenzene-d5                                  0210555
             10   117  659   26.6223   10     97301.UG/L     50.0000   02240C3
            C230     Toluene                                           0236F52
             11    92  629   25.4587   10     97525.UG/L     52.6810   0243ADC
            CS05     Toluene-8                                         0254B5F
             12    98  624   25.2646   10    131454.UG/L     49.5044   02658C4
            C235     Chlorobenzene                                     0274133
             13   112  663   26.7776   10    107967.UG/L     49.8718   02870D4
            CS10     Bromofluorobenzene                                0297E06
             14    95  812   32.5565   10     61665.UG/L     49.8151   03049CA
ENDFILE                                                                0317BE0
```

APPENDIX 2

Preparing Quality Assurance Project Plans

INTRODUCTION

This guidance document on the preparation of a Quality Assurance Project Plan (QAPjP) has been developed to assist field project and laboratory managers in developing data quality objectives for their projects. A Work Plan (WP) should be prepared for all work being performed. The WP consists of the Quality Assurance Program Plan (QAPP), Sampling and Analysis Plan (SAP), Health and Safety Plan, and other regulatory required documents.

The SAP consists of a Field Sampling Plan (FSP), which details the field activities, and a QAPjP, which details the necessary steps that ensure data collected during field activities meet objectives outlined in the WP and the FSP.

Sixteen sections are required for a QAPjP. These sections are described in "Interim Guidelines and Specifications for Preparing Quality Assurance Project Plans" (EPA QAMS-005/80) and the "Preparation Aids for the Development of Category 1 Quality Assurance Project Plans" (EPA/600/8-91/003). This document presents a generic overview of each of the sixteen sections of the QAPjP.

SECTION 1 TITLE PAGE AND SIGNATURE PAGE

1.1 Generic

The QA Project Plan must be signed by key project personnel as well as key personnel from any subcontractors. These signatures indicate that the key personnel have read the appropriate sections of the QA Project Plan and are committed to implementing the plan.

The QAPjP Title Page will address each of the following items:

- Title
- Date
- Revision number
- Issuing agency

For document control format the Section number, revision, date, and page should be recorded in an upper corner of each page. This format requirement is illustrated below.

Section No.: 1.0
Revision: 0
Date: October 45, 2005
Page: 3 of 5

The QAPjP Signature Page will contain the signatures of the project manager and immediate supervisor, project and program QA official, and agency's project and QA official. The following presents a title and signature page for a laboratory QAPjP.

LABORATORY QUALITY ASSURANCE PROJECT PLAN

for

CLIENT
SITE NAME
ADDRESS
DATE

____

CLIENT NAME Project Manager

____

CLIENT NAME Project QA Officer

____

LABORATORY Program Administrator

____

LABORATORY QA Director

____

EPA Region # QA Officer

SECTION 2 TABLE OF CONTENTS

2.1 Generic

The QAPjP Table of Contents will address each of the following items:

Introduction
Serial listing of each of the QAPjP components

A listing of any appendices which are required to augment the QAPjP.
The following lists the table of contents for the generic table of contents.

TABLE OF CONTENTS

1. TITLE PAGE
2. TABLE OF CONTENTS
3. PROJECT DESCRIPTION
4. PROJECT ORGANIZATION AND RESPONSIBILITIES
5. QUALITY ASSURANCE OBJECTIVES
6. SAMPLING PROCEDURES
7. SAMPLE CUSTODY
8. CALIBRATION AND ANALYTICAL PROCEDURES
9. DATA REDUCTION, VALIDATION AND REPORTING
10. INTERNAL QUALITY CONTROL CHECKS
11. PERFORMANCE AND SYSTEM AUDITS
12. PREVENTATIVE MAINTENANCE
13. PROCEDURES TO ASSESS DATA QUALITY INDICATORS
14. CORRECTIVE ACTION
15. QUALITY ASSURANCE REPORTS TO MANAGEMENT
16. REFERENCES

SECTION 3 PROJECT DESCRIPTION

This section presents the necessary elements of a project description.

3.1 Generic Overview Project Description

This section provides the process or environmental system that is to be tested, the project objectives, a summary of the experimental design, and the proposed project schedule. The section typically contains the subsections listed below, plus others as needed.

3.1.1 General Overview

This section provides a brief synopsis of the overall project. In one or two paragraphs the programmatic and regulatory settings are described. The purpose(s) of the study should be described, along with the decisions that are to be made and the hypothesis to be tested. Other anticipated uses of the data should also be noted. The type of process or environmental system that is to be tested should also be described briefly.

If the project is preceded by various preliminary investigations of the site or process, the results of such preliminary tests can be included in this section, with details presented in an appendix.

Often the project must demonstrate compliance with applicable regulations and such regulations should be summarized if they are applicable to the project.

3.1.2 The Process, Site, Facility, or System

In this section describe the process, site, facility, or environmental system that will be tested. Include flow diagrams, maps, charts, etc., as needed. Approximate mass or volumetric flow rates should be indicated to permit proper evaluation of the sampling and process monitoring procedures. Additional diagrams are often included to unambiguously describe sampling points. Such detailed diagrams can be included either in this section or in Section 4 (Site Selection and Sampling Procedures).

This section should contain enough material to permit a technical reviewer who is unfamiliar with the project to assess the proposed sampling strategy.

3.1.3 Statement of Project Objectives

Projects typically have multiple objectives and those project objectives should be summarized and stated clearly in this section. Project objectives should be stated in numerical terms whenever possible. Avoid scattering statements of project objectives throughout the sampling and analytical sections of the QA Project Plan, or among several documents.

Some objectives cannot be stated in entirely quantitative terms. For example, one objective might be to determine the dominant chemical species in a reactor. Or, for a test of a new incinerator design, one might want to determine the range of combustion conditions that result in a molten ("slagging") ash. Clear project objectives form the basis for designing the sampling, analysis, and quality control strategies. Thus, it is essential that both the decision makers and technical staff understand and agree with these objectives.

3.1.4 Experimental Design

List all measurements that will be made during the project. It may be possible to classify them as critical or noncritical measurements. Critical measurements include those that are necessary to achieve project objectives; they may include either on-site process measurements or chemical measurements. Noncritical measurements are those used for process control or background information.

It is common practice to test pollution control equipment or processes in different operating modes or stages. For example, an incinerator might be operated under warm-up conditions followed by incineration tests at three different feed rates. The various test conditions should be summarized in this section. Indicate the length of time anticipated for each test condition. When appropriate, indicate how equilibrium conditions will be established before sample collection begins.

A summary should be prepared for each planned sample in tabular form for all measurements. Ideally, this table should indicate the total number of samples for each sample point, including quality control and reserve samples. Relate the individual samples to the sampling points shown in the process diagram, if applicable. For projects involving a large number of samples or analyses, it may not be possible to include all QC and reserve samples in a single table. For such cases, two or more tables may be necessary, the first to summarize the primary (non-QC) samples, and the others to show which QC samples are associated with which analyses. Information on the total number of all sample types is needed both to perform a comprehensive review and to estimate analytical costs.

In this section include enough information to permit a technical person unfamiliar with your project to evaluate the sampling and analytical approach. Avoid repeating material from a sampling or analytical plan. Complete a necessary description once, then cite it in all other documents. Most important, essential details are readily overlooked when buried in repetitious or overly lengthy documentation.

Cite applicable regulations, if any.

3.1.5 Schedule

Indicate start-up and ending dates, including those for preliminary studies and field and laboratory activities.

3.2 Project Specific Description

Provide additional details regarding the project that may not be covered in the generic description. This will include some or all of the following information.

3.2.1 Site Description

Give the site location in latitude and longitude. Indicate specific terrain features (e.g., proximity to a stream, creek, lake, etc., distance to the nearest town, elevation of the site, distance to the nearest homes, whether the site is graded, the size of the site or operable unit, attach maps and drawings that show location in relation to roads and topography).

3.2.2 Site History

Provides a complete history of the site. A table with a chronology of events is included.

3.2.3 Hydrogeological Characteristics

Present a complete hydrogeological description of the site and the surrounding area with groundwater wells fully detailed.

3.2.4 Population and Environmental Resources

The surrounding population is described and any significant environmental resources are detailed.

3.2.5 Intended Data Uses

State the decisions that will be made based on the data if you know them. The following generic statements will apply to a number of sampling and analysis projects.

The data from each field and laboratory test will be used to address the specified objectives. The sampling of surface water, sediment, vegetation, and fauna will be used to assess whether site-generated contaminants have migrated off the site and to what extent they are present in the biota. Data from the monitoring wells will indicate the extent to which the groundwater may have been contaminated and in what direction it may be migrating. Geophysical results will aid in determining the extent of groundwater contamination. Data from soil samples will be used to assess the nature and extent of soil contamination at the site as well as determine the presence of any hot spots.

The results of the investigation will be used to assess the potential risks that contamination presents to human health and the environment and to evaluate feasible alternatives to mitigate any risks that are identified. The data generated during field activities and by the laboratories must be sensitive enough for meaningful use of the data. In addition, the data must be sufficiently validated to stand up to the scrutiny of potential enforcement activities.

3.2.6 Previous Sampling Results

Describe in detail with charts and figures previous sampling results to the extent they are known.

3.3 Laboratory Project Description

This quality assurance project plan (QAPjP) outlines the specific quality assurance procedures to be followed by *THE LABORATORY* in generating chemical analyses related to the activities of the *XYZ COMPANY*. This plan calls for the analysis of groundwater, soil, air, sludge, and drinking water (etc).

The purpose of this QAPjP is to provide a detailed description of all the elements involved in the generation of data of acceptable quality and completeness for the samples collected for the XYZ COMPANY. The data must be comprehensive enough to evaluate all the project objectives and must be sufficiently validated to stand up to the scrutiny of potential enforcement activities.

3.3.1 Purpose

The purpose of this QAPjP is to provide a detailed description of all the elements involved in the generation of data of acceptable quality and completeness for the monitoring of volatiles, semivolatiles, metals, pesticides, PCBs, and MMMM chemical (LIST ALL CHEMICALS, MATERIALS, AND SPECIAL ANALYSES).

3.3.2 Scope

The scope of this QAPjP is to outline the QC requirements for all data generated during the project based on quality judgements using the following three types of information:

1. Data that demonstrate overall laboratory capabilities, including internal and external performance and system audits to ensure that there are adequate facilities and equipment, qualified personnel, documented laboratory procedures, accurate data reduction, proper validation, and complete reporting.
2. Data that measure the daily performance of the laboratory according to the specific method employed. This includes data on calibration procedures and instrument performance.
3. Data that evaluate the overall quality of the package that is used to determine precision, accuracy, representativeness, completeness, and comparability which is in compliance with the data quality objectives listed in Section 5. Such data includes but is not limited to laboratory method blanks and duplicate control samples.

3.3.3 Analyses

The samples (groundwater, soil, air, surface water, sediment, and fauna) will be analyzed for the following parameters. List every parameter to be tested.

SECTION 4 PROJECT ORGANIZATION AND RESPONSIBILITIES

4.1 Generic Project Organization and Responsibilities

This section demonstrates that the project organization is adequate to accomplish project goals, and that all responsibilities have been assigned.

Provide a table or chart illustrating project organization and lines of authority. For each organizational entity or subcontractor, identify by name all key personnel and give their geographic locations and phone numbers. The organizational chart should also include all subcontractors and their key points of contact. Separate organizational charts for subcontractors may be needed. The organizational chart should identify QA managers, including those of subcontractors, and should illustrate their relationship to other project personnel. The QA Managers should be organizationally independent of the project management so that the risk of conflict of interest is minimized.

Describe the responsibilities of all project participants, including QA managers. Be sure to indicate responsibility for each type of analysis, physical measurement, and process measurement. This summary should designate responsibility for planning, coordination, sample collection, sample custody, analysis, review, and report preparation.

Any relevant certifications (e.g., OSHA 40-h course for Hazardous Waste Site Operations) held by members of the project team should be noted. In some cases, these certifications augment relevant educational and work experience. In other cases, certification may be required by law.

Describe the frequency and mechanisms of communications among the contractor, the contractor's QA manager, and the laboratory project manager, as well as among contractor and subcontractors. Give the regular schedules for progress reports, site visits, and teleconferences, and describe special occurrences that trigger additional communication.

Communication is a key element in achieving project goals. Because success largely depends on the prime contractor, effective monitoring of all project activities, including those of subcontractors, is vital. Procedures for communicating and monitoring all levels of activity must be clearly delineated.

Be sure to:

- Demonstrate that your QA manager is independent of project organization.
- Provide names, locations, organizational affiliations, and telephone numbers for key personnel

SECTION 5 DATA QUALITY OBJECTIVES

This section presents the necessary elements in the determination of quality assurance objectives in terms of precision, accuracy, completeness, and comparability. A generic overview of these QA objectives is presented.

5.1 Generic Quality Assurance Objectives

Quality assurance objectives are specifications that measurements must meet in order to achieve project objectives. For example, in ordering a pump for a chemical plant, factors such as capacity, pressure, and materials of constructions must be specified. Similarly, precision, accuracy, detection limits, and completeness must be specified for physical/chemical measurements. These specifications may change depending on concentration. For example, if a decision point is at a concentration (e.g., 5 mg/l) then the precision and accuracy requirements do not have to be as stringent for samples that are above, say 50 mg/l. Likewise, these requirements may be relaxed for samples with concentrations below 0.5 mg/l.

Additional analytical requirements are described qualitatively in terms of representativeness and comparability. Quality assurance objectives are needed for all critical measurements (see Section 3.2) and for each type of sample matrix (soil, water, biota, etc.). Included are quality assurance objectives for physical as well as chemical measurements.

Once these QA objectives are set, the measurement systems are then designed to meet them. The project manager, analytical chemists, and other principals must agree on the feasibility and appropriateness of these objectives.

5.1.1 Determining QA Objectives

QA objectives must be defined in terms of project requirements, and not in terms of the capabilities of the intended test methods. Of course, the QA objectives must be achievable by available methods, and for this reason it is important that the laboratory review the QA objectives. When QA objectives exceed the capabilities of available methods, either the methods must be modified or the test plan must compensate for the deficiencies. Modifications may often be as simple as collecting a larger sample. If nonstandard or significantly modified test methods are required, however, the QA Project Plan should include laboratory validation data (or at least a description of how these data will be generated), to prove that the method is capable of achieving the desired performance.

As part of the quality objectives process, quantitative data QA objectives, based upon project requirements, should be implemented. At a minimum, these quantitative data QA objectives should be specified for accuracy, precision, completeness, and method detection limit.

5.1.2 Data QA Objectives: Precision, Accuracy, Method Detection Limit, and Completeness

QA data objectives for precision, accuracy, method detection limit, and completeness should be presented in a QA objectives table. Be sure to include QA objectives for all matrix types and to indicate the units in which these QA objectives are given. Summary tables are very helpful to the laboratory which must meet these objectives.

Because precision, accuracy, and completeness can be measured in various ways, explain the method to be used by the laboratory. Analytical precision of the data is the difference between the results of analysis of duplicate samples relative to the average of those results for a given analyte and is expressed as the relative percent difference (RPD). Analytical accuracy, expressed as a percent, is the recovery of a standard material or an analyte that has been added to the sample at a known concentration before analysis. Completeness is a measure of the relative number of analytical data points that meet all the acceptance criteria for data.

If precision, for instance, is to be determined by duplicates, explain whether sample splitting will occur in the laboratory, during sampling, or at some other stage. Then summarize all information in either text or tabular format.

The following statements are examples of descriptions for precision, accuracy, method detection limits, and completeness:

- Precision objectives for all the listed methods except pH are presented as relative percent difference (RPD) of field duplicates. Precision objectives for pH are listed in pH units and expressed as limits for field duplicates.
- Precision objectives for unconfined compressive strength are given as relative standard deviation for triplicate sets.
- Accuracy objectives for organic compounds and metals are given as a percent recovery range of laboratory matrix spikes. Accuracy objectives for temperature measurements are absolute deviations in °C.
- Detection limits are defined as the method detection limit (MDL) multiplied by the dilution factor required to analyze the sample. MDLs are typically estimated by determining the concentration in seven identical samples (spiked at a concentration near what you believe is slightly above the MDL). For example, if you think the MDL is near 1 μg/l then you should spike the samples at about 5 to 10 μg/l. The standard deviation from the seven measurements is used to establish the MDL. For example, if the standard deviation for the samples is 3 μg/l then the MDL (95% confidence or two sigma) is about 6 μg/l. At 99% confidence, or three sigma, it would be 9 μg/l. This means that if you measure a concentration at or just above 9 μg/l then you can be 99% sure the concentration is not zero. Obviously, higher sample concentrations carry even higher levels of confidence.
- Completeness is defined as the number of measurements judged valid compared to the number of measurements needed to achieve a specified level of confidence in decision making. For example, you might estimate that ten valid measurements should suffice to demonstrate that a chemical concentration in a discharge is less than 50 μg/l at a confidence of 90%. In other words, ten samples will provide enough precision to say whether the discharge is above or below 50 μg/l with 90% confidence. To allow for a margin of error in the estimated number of samples, 15 valid measurements are planned, resulting in a completeness objective of 150%. An additional six spare samples will be collected but not analyzed, unless needed to achieve the desired precision/confidence level.
- When analyzing for a large number of organic compounds by GC or GC/MS, it is usually unnecessary to list QA objectives for each compound separately. Instead, list QA objectives according to compound type or class. In other cases, for example, where detection limits are derived from applicable regulations, list detection limits for individual compounds as needed.

The QA Project Plan must explain how the QA objectives are to be interpreted in a statistical sense. QA objectives are often interpreted in a sense that all data must fall within these goals; for such projects any data that fail to satisfy the QA objectives are rejected and corrective action is undertaken. However, other interpretations are possible. For example, the project requirements may be satisfied if the average recovery is within the objectives; that is, individual excursions beyond the objectives might be permitted, but the average recovery would have to satisfy the goals described in this table. Whatever the case, it is important to describe in this section how tabulated QA objectives will be interpreted.

5.1.3 Data QA Objectives; Comparability and Representativeness

Comparability is the degree of confidence to which one data set can be compared to another. Data comparability is achieved through the use of standard sampling and analytical techniques. Comparability is achieved by the use of consistent methods and by traceability of standards to a reliable source.

Representativeness is the degree to which a sample or group of samples accurately and precisely represents a characteristic of a population, parameter variations at a sampling point, a process condition, or an environmental condition. An environmental sample is representative for a parameter of interest when the average value obtained from a group of such samples tends toward the true value of that parameter in the actual environment, as the number of representative samples is increased. Representativeness is normally achieved by collecting a sufficiently large number of unbiased samples. Statistical methods are used in the sampling design to ensure that the appropriate numbers of samples are taken.

The QA Project Plan should demonstrate that adequate comparability and representativeness will be achieved as required by the project.

5.1.4 Other QA Objectives

Some projects may require additional QA objectives, such as mass balances. Requirements for all additional QA measurements should be stated in this section.

5.1.5 What if DQOs Are Not Met?

This section should include a discussion of the impact of not meeting one or more QA objectives. Will the project be a complete loss? Will some, but not all of the project goals still be realized? Will the statistical confidence level be reduced? Are there legal or regulatory ramifications? Answers to such questions help provide a critical perspective on the QA Program.

5.2 Specific Quality Assurance Objectives

5.2.1 Data Requirements

Data requirements might be as follows. Analytical sampling results will be used to determine if contaminant concentrations exceed any applicable or relevant and appropriate requirements (ARARs), to develop a conceptual risk assessment model of the site, and for remediation and feasibility studies. Data gathering during the RI will be used to identify cost effective, environmentally sound, long-term measures for remediation of the site.

A work plan has been developed to evaluate the site as completely, efficiently, and practically as possible. The work plan includes sampling procedures, analytical methods, and special analyses for quantitatively assessing the site. The investigation process contains quality assurance goals and QA procedures to measure data quality. To ensure that the data quality goals are met, certain DQOs are established for the data to be gathered during the project.

5.2.2 Data Quality Objectives and Quality Assurance Objectives

To measure and control the quality of analysis and to ensure that the DQOs are met, certain QA parameters are defined and utilized in data analysis activities in this project. They are defined as follows:

- **Precision** — Precision measures the reproducibility of measurements under a given set of conditions. Specifically, it is a measurement of the variability of a group of measurements compared to their average value. As such, it is extremely important to have a precise analysis. In this project, precision will be determined by the EPA-approved analytical methods and measured in terms of relative percent difference (RPD). For USEPA CLP analyses, the RPD control limits are listed in the appropriate contract SOW.
- **Accuracy** — Accuracy measures the bias in a measurement system. Sources of error include the sampling process, field contamination, preservation, handling, shipping, sample matrix, sample preparation, and analysis techniques. In this project, sampling accuracy will be evaluated by comparison of analytical results to field duplicates. Analytical accuracy will be assessed by the analysis of matrix spike recoveries.
- **Comparability** — Comparability expresses the confidence with which one data set can be compared with another. By using standard, established sampling, analytical and reporting procedures, this SAP allows the comparability of all data generated in this project with previous data, historical data bases, and data that may be required in later phases.
- **Completeness** — The completeness objectives for this project will require that at a minimum, 80% of the QA/QC data meet the Quality Assurance Objectives for this project as stated in this section.
- **Representativeness** — The representativeness of samples will be assured by the collection procedures outlined by the Field Sampling Plan and the Standard Operating Procedures (SOPs) included in this QAPP. Areal and depth representatives will be assured through procedures outlined in the Work Plan.
- **Sensitivity** — The data generated during this project will be sensitive enough to meet ARAR criteria. CLP contract-required quantitation limits (CRQLs) will be used in general. In instances for which RAS CRQLs are too high to allow comparison to ARARs, CLP SAS with low detection limits will be used.

5.2.3 Data Quality Objectives Development

In this section the data quality objectives for each data collection activity are described along with the necessary QA/QC requirements.

5.2.3.1 Groundwater

Groundwater will be sampled and analyzed to characterize the nature and extent of ground-water contamination and to assess the risk any contamination may pose to public health and the environment. The data will be used to identify the location of any groundwater contamination, to aid in determining contaminant source locations, to determine if contaminant migration poses any risk to public water supply aquifers, and determine if any ARARs have been exceeded. The data will be used to characterize the site. The first round of ground-water samples will be analyzed according to CLP procedures. Based on the review of validated data from the first round of sampling, selected samples from the second round of sampling will be analyzed with low quantitation limits in order to compare contaminant concentrations to ARARs. Samples for analysis with low quantitation limits will be selected in consultation with the appropriate agency. The data quality objective for the first round of sampling and a portion of the second round samples will be characterized by rigorous QA/QC protocols and documentation.

Add specific data quality objectives for groundwater here.

5.2.3.2 Soil Boring and Surface Soil Investigations

The objective of the soil boring and soil sampling program is to define the nature and extent of the contamination in the vadose zone. Samples will be collected in conjunction with the installation of groundwater monitoring wells and where surface soils exhibit staining.

Data from this investigation will be used to characterize the site and to determine the risk to the public health and the environment associated with the contamination at this site. The specific pathway of concern is groundwater contamination which could affect residential water supply wells in the vicinity of the site.

5.2.3.3 Surface Water and Leachate Sample

The objectives of the surface water and leachate sampling is to determine if site-generated contaminants have been, or are being, transported to open areas downslope of the site. Data from this investigation will be used to characterize the site and to determine if the migration of contaminants from the site have affected the growth and production of indigenous flora and fauna. Also, analytical results will be used to evaluate the risk to public health from exposure to surface water.

5.2.3.4 Sediment Sampling

The purpose of the sediment sampling is to determine if site-generated contaminants are present in pond and creek sediments at concentrations which have the potential to affect growth and production of indigenous flora and fauna.

5.3 Laboratory Quality Assurance Objectives

This section presents the data quality assurance objectives that have been agreed to by the sample-submitting client and the laboratory.

5.3.1 Quality Assurance Objectives

Quality assurance objectives can be expressed in terms of precision, accuracy, representativeness, comparability, and completeness.

Adherence to the DQOs will be quantitatively measured by comparing the results of the Duplicate Control Samples (DCS) to control limits. DCS consist of a standard control matrix that is spiked with a group of target compounds representative of the method analytes. A DCS pair is analyzed for every 20 samples processed by a method.

Samples are analyzed in lots of less than 20, due to holding time or turn-around time requirements. Since it is necessary to have a measure of laboratory performance with each batch of samples a single control sample (SCS) is processed. An SCS consists of a control matrix that is spiked with surrogate compounds appropriate to the method being used. In cases where no surrogate is available (e.g., metals or wet chemistry) a single DCS serves as the control sample. An SCS is prepared for each sample lot for which the DCS pair is not analyzed.

The DCS pair is used to monitor both the precision and accuracy of the analytical method on an ongoing basis, independent of matrix effects. DCS are monitored for accuracy (relative percent recovery) of each analyte and precision (relative percent difference — RPD) between each analyte in the DCS pair. Section 11.3 defines the Laboratory Calculation of Data Quality Indicators.

Percent completeness is defined as the number of valid data points obtained divided by the number of data points attempted. To be considered complete, the data set must contain all QC check analyses verifying precision and accuracy for all analytical protocols. Less obvious is whether that data are sufficient to achieve the goals of the project. All data are reviewed in terms of goals in order to determine if the data base is sufficient.

Representativeness can be defined as the degree to which the data accurately represent the media from which it is collected. Representativeness can be measured by comparison of field duplicate results. Comparability expresses the confidence with which two data sets can be compared. Comparability can be measured by the adherence to the QC practices and criteria contained in this plan.

5.3.2 Control Limits

Control limits are generated for the standard control matrix based upon historical data. Control limits for accuracy and precision are subject to periodic updating. The control limits used will be those in effect at the time the samples are analyzed.

5.3.3 Duplicate Control Samples

Precision and accuracy are assessed by the laboratory by comparing the results of DCS to the control limits. Accuracy is expressed as the average percent recovery of the DCS pair and precision is expressed as the relative percent difference.

For all *inorganic* tests, if DCS are out of control limits, all the samples which are associated with the unacceptable DCS must be reprepared and/or reanalyzed. For multianalyte *organic* tests, if greater than 20% of the accuracy or precision results for the DCS are out of control, the data are considered suspect and the samples associated with the unacceptable DCS are reprepared and/or reanalyzed. If less than 20% of the accuracy or precision results for the DCS are out of control, the data are investigated and reported if the data meet the QC requirements of the method.

5.3.4 Single Control Samples

Recovery data generated from the SCS are compared to control limits that are established for each of the compounds being monitored. Analytical data that are within the control limits are judged to be in control. Data that are outside of control limits are considered suspect and corrective action must be performed, see Section 12.3. The associated SCS data are reported with each set of sample results to enable a quality assessment of the data.

5.3.5 Matrix-Specific QC

Matrix-specific QC samples are analyzed when requested on the chain of custody. The performance of matrix specific analyses for aqueous matrices requires additional sample volume which must be collected and submitted at the same time as the original routine sample.

Typical frequencies include one pair of matrix spikes for each sample type per batch of 20 or fewer samples collected on one day. However, frequencies of collection may vary based on the DQOs of the project.

For organic analysis, the percent recovery and the relative percent difference (RPD) of the matrix spike (MS) pair will be calculated. For inorganic analyses, the MS percent recovery and matrix duplicate RPD will be calculated. This allows for demonstration of the effect of the matrix on the method performed. Reextraction and reanalysis decisions are made based on the DCS, Method Blanks, and QC requirements of the methods.

5.3.6 Surrogates

Surrogates are organic compounds which are similar to the analytes of interest in chemical behavior, but are not normally found in environmental samples. Surrogates are added to samples to monitor the effect of the matrix on the accuracy of the analysis. Results are reported in terms of percent recovery. Surrogate recoveries for GC/MS analyses for samples are compared with the limits published in the EPA CLP SOW.

5.3.7 Method Blanks

Method blanks, also known as reagent, analytical, or preparation blanks, are analyzed to assess the level of background interference or contamination that exists in the analytical system and which might lead to the reporting of elevated concentration levels or false-positive data.

As part of the laboratory program a method blank is analyzed with every batch of samples processed. A method blank consists of reagents specific to the method which are carried through every aspect of the procedure, including preparation, clean-up, and analysis. The results of the method blank are evaluated, in conjunction with other QC information, to determine the acceptability of the data generated for that batch of samples.

Ideally, the concentration of target analytes in the blank should be below the Reporting Limit for that analyte. In practice, however, some common laboratory solvents and metals are difficult to eliminate to the ppb levels commonly required in environmental analyses. Therefore, criteria for determining blank acceptability must be based on consideration of the analytical techniques used, analytes reported, and reporting limits required.

For organic analyses, the concentration of target analytes in the blank must be below the reporting limit for that analyte in order for the blank to be considered acceptable. An exception is made for common laboratory contaminants (methylene chloride, acetone, 2-butanone, and phthalate esters) which may be present in the blank at up to five times the reporting limit and still be considered acceptable.

For nonroutine organic analyses, other components may be established as common contaminants for that particular analyses. If, upon thorough review of the method during validation it is deemed impossible to eliminate trace amounts of analytes from the process, these analytes are likewise allowed at up to five times the reporting limit.

For metals and wet chemistry analyses, where the reporting limits are typically near the instrument detection limit (IDL), the policy is that the concentration of the target analytes in the blank must be below two times the reporting limit. If the blank value lies between the reporting limit and two times the reporting limit, the analytes in the associated samples are flagged to indicate contamination was present in the blank. A blank containing an analyte(s) above two times the reporting limit is considered unacceptable unless the lowest concentration of the analyte in the associated samples is at least ten times the blank concentration or the concentration of the analyte in all samples associated with the blank is below the reporting limit.

In addition, for wet chemistry tests, the method SOP directs how the blank is treated. Generally, a reagent blank is used both to zero the equipment and as one of the calibration standards. If a preparation step is required for the analysis then a prep blank is also analyzed to determine the extent of contamination or background interference. Blanks have no application or significance for some wet chemistry parameters (e.g., pH).

If the blank does not meet acceptance criteria, the source of contamination must be investigated and appropriate action taken and documented. Investigation includes an evaluation of the data to determine the extent and effect of the contamination on the sample results. Corrective actions may include reanalysis of the blank, and/or repreparation and reanalysis of the blank and associated samples.

For organic and metals analyses, and selected wet chemistry tests, method blank results are reported with each set of sample results. Sample results are not corrected for blank contamination *unless* required by the analytical method or requested by the client. Occasionally, the laboratory may report data associated with an unacceptable blank due to time or materials constraints. In these cases the actual observed value is reported in the method blank and all sample results are flagged to indicate contamination was present in the associated method blank.

SECTION 6 SAMPLING PROCEDURES

This section must describe the sampling procedures which are responsive to those project objectives stated in Section 1.1.2.

This section explains the overall sampling strategy and the specific sampling procedures that will be employed.

6.1 Generic Sampling Procedures

This section describes the specific procedures that will be used for collecting and preserving samples. It is vital that each of the following points be completely and clearly addressed to ensure that activities undertaken are adequate to meet project objectives.

Discuss each sampling procedure that will be employed. For approved procedures, a reference is sufficient. Other sampling procedures should be summarized in the text, and additional details should be provided in an Appendix. (Copies of ASTM sampling procedures are frequently appended.)

- Prepare a list of analytes, sample volumes to be collected, and the amount of sample that is required for each analysis. Note whether the required amount is intended for matrix spike/matrix spike duplicate terminations or for a single determination only. Be sure to have your laboratory manager review this table to ensure that the sample volume or mass is sufficient for all intended analyses.
- Describe any compositing or sample splitting procedures that will be employed in the field or laboratory. Note whether the host facility requires additional split samples for its own analyses.
- Describe any sampling equipment that will be used, and how this equipment will be calibrated.
- Explain how sample containers will be cleaned to prevent sample contamination, and how new sample containers will be checked for contaminants.
- Describe the numbering sequence which ensures that each sample will be assigned a unique number.
- Describe the containers used for sample collection, transport, and storage for each sample type. Include sample preservation methods, noting specific reagents, equipment, supplies, etc., required for sample preservation, and the specific time requirements for shipping samples to the laboratory. Note refrigeration conditions and holding times that will be employed.
- Describe the procedures used to record sample history, sampling conditions, and any other pertinent information; include examples of forms that will be employed. In nonerasable waterproof ink, record all aspects of sample collection — field data and observations, problems encountered, and action taken to resolve the problems — in a bound notebook with consecutively numbered pages. Pages should never be removed from this logbook.
- Include an example of the sample label to be used.

SECTION 7 SAMPLE CUSTODY

This section of the QAPjP describes:

- How sample custody will be maintained and recorded.
- Procedures that will be used to maintain chain of custody during transfer from the field to the laboratory, within the laboratory, and among contractors and subcontractors.

- Provisions for the archiving of all shipping documents and other paperwork received at the laboratory with the samples.
- Procedures that will ensure sample security. For example, samples should always be stored in locked refrigerators.
- Examples of forms that will be used to maintain sample custody in the field, during shipping, and in the laboratory.
- Procedures for within-laboratory chain of custody. Such procedures should allow for unambiguous tracing of the samples through the laboratory. This "paper trail" should include a record of the individuals responsible for custody of samples, extracts, digests, etc., at all times in the laboratory. Finally, disposal or consumption of samples should be documented.

Insert copies of chain of custody forms, seals, etc. which you routinely use here.

Certain information is required in this section — such as sample custody procedures, forms, labels, and sampling methods. Typically, a sampling team or a laboratory will use the same chain of custody procedures from one project to the next. Such procedures may be written as SOPs, which can then be appended to subsequent QAPjPs. Be sure to use document control format in the SOPs to alert the reviewer to any recent changes in the procedure.

7.1 Generic Sampling Procedures

Occasionally samples are spilled, contaminated, accidentally evaporated to dryness, or otherwise compromised before or during sampling and analysis. Sample custody allows detection of such problems should they occur and minimizes such occurrences by assigning responsibility for all stages of sample handling. Sample custody procedures require that the possession and handling of the sample from the moment of its collection through analysis be documented by written record. Sample custody is maintained when the samples are in a secure area, or in the view of, or under the control of, a particular individual. The records of everyone handling samples are maintained so that a sample history can be reconstructed later, should the need arise.

Record-keeping documentation includes using indelible blue or black ink, field notebooks or approved forms, and waterproof labels. The chain of custody record lists:

- The names of all sample custodians in the field and in each of the laboratories
- Sample identifier
- Sampling date
- Sample matrix
- Number of containers
- Analysis requested and turn-around time required

7.2 Sample Custody

7.2.1 Overview

Sample custody procedures ensure the timely, correct, and complete analysis of each sample for all parameters requested. Sample custody documentation provides a written record of sample collection and analysis, and is required evidence in any enforcement actions against potentially responsible parties (PRPs). The sample custody procedures provide for specific identification of samples associated with an exact location, the recording of pertinent information associated with the sample, the time of sample collection and any preservation techniques, and provide for a written chain of custody record which serves as physical evidence of sample custody. Custody procedures will adhere to the procedures outlined in the USEPA document "Users Guide to the Contract Laboratory Program", December 1988. The chain of custody documentation system provides the means to individually identify, track, and monitor each sample from the time of collection through final data reporting. Sample custody procedures are developed for four areas which are described below.

7.2.2 Field Records and Sample Collection

Chain-of-custody procedures document pertinent sampling data and all transfers of custody until the sample reaches the analytical laboratory. The chain-of-custody procedures assure that all samples are uniquely identified, that the correct samples are analyzed and are traceable to their source. The documentation and custody requirements for field monitoring and samples collected for field analysis are different from samples collected for laboratory analysis.

7.2.3 Field Records

The field logbook serves as the permanent record of all analyses conducted in the field. The field logbook will be a bound notebook with all entries recorded in pen. All in-field analytical results are recorded in the field logbook. Each sample or test location is uniquely and clearly identified according to the location on the site or on the predetermined grid pattern. The field technicians will record or reference the following information in the field logbook:

1. Field analytical equipment (include serial number)
2. Other measuring equipment (include serial number)
3. Calculations
4. Results
5. Calibration data for equipment

Observations such as sample matrix, sampling conditions, or any problems will also be recorded. The field technicians will sign and date each field logbook page.

7.2.4 Sample Collection Documentation Procedure

All samples collected will have a written record documenting and describing the sampling procedures in the sampling logbook. The following will be recorded in the sampling logbook for each sample collected:

1. Sample ID number
2. Sample location
3. Sample collection equipment used
4. Field analytical equipment used
5. Field measurement results
6. Number and types of sample container used
7. Sample matrix
8. Name of sample collector(s)

A description of the sample, including sample matrix and observations made during sampling, will be included in this logbook. In addition, notes on the representativeness of the sample, sample phases, compositing, and mixing of sample contents will be recorded in the sampling notebook.

7.2.5 Sample Identification

The documentation system for laboratory samples is based on the sample documentation system described in the "Users Guide to the Contract Laboratory Program" (USEPA, 1988).

All samples collected will have a label that contains the following information:

1. Project number
2. Field ID or sample station number
3. Date and time of collection
4. Designation of sample as grab or composite
5. Type of sample
6. Description of sample location
7. Sample preservation notes
8. Analytical parameters

CLP samples will have a sample tag affixed to each container with the same information. CLP samples will have a specific CLP label. Non-CLP samples will have a Malcolm Pirnie ARCS project label. All sample tags and labels will be completed in waterproof ink by the sampler. The sample labels will be taped over with clear tape before shipping.

The project number is used instead of the site name in order to preserve the anonymity of the site. The bottles will be prenumbered according to the Field Sampling Plan's numbering scheme.

7.2.6 Sample Traffic Reports

Sample traffic reports will be filled out for all CLP-RAS analyses in accordance with the procedures given in the "Users Guide to the CLP", USEPA, 1988. Organic and inorganic sample traffic reports are provided to Appendix A.

7.2.7 Chain of Custody Procedures

At the time of sampling, a chain-of-custody form shall be completed for each sample or group of samples. The form provides both a chain-of-custody record and sample analysis request form. Each sample will be recorded on this form. The following information will be recorded on the chain-of-custody/sample request form:

1. Project name
2. Signature of samplers
3. Sampling station number
4. Date and time of collection
5. Grab or composite sample designation
6. Sample matrix with brief description
7. Sampling location description
8. Field identification number
9. Analyses requested
10. Preservation technique
11. Signatures and dates for transfers of custody
12. Air express/shipper's bill of lading identification numbers

The chain-of-custody form/sample analyses request sheet serves as an official communication to the laboratory of the particular analyses required for each sample. The chain-of-custody record will accompany the samples from the time of sampling through all transfers of custody, and be kept on file at the laboratory where samples are analyzed and archived. The form is filled out in triplicate, one copy is retained by the site manager, one is sent to the laboratory, and the third will be submitted to the overall project manager. The sampler completes a chain-of-custody record to accompany each shipment from the field to the laboratory. Separate chain-of-custody records are filled out for split samples. Errors must be crossed through with a single line, initialed, and dated. The completed chain-of-custody record is put in a zipper-lock bag and taped to the inside cover of the sample shipping container. The container is then sealed with custody seals and custody is transferred to the laboratory.

7.2.8 Transfer of Custody and Shipment

The custody of the samples must be maintained from the time of sampling, through shipment and relinquishment to the laboratory. Samples required to be analyzed by the Contract Laboratory Program will be shipped according to procedure in the "Users Guide to the Contract Laboratory Program". Instructions for transferring custody are given below.

1. Samples are accompanied by a chain-of-custody record. When transferring the custody of samples, the individuals relinquishing and receiving will sign, date, and note the time on the record. This record documents sample custody transfer from the sampler, through the shipper, to the analytical laboratory. A shipper will sign the record. A common carrier will usually not accept the responsibility for handling chain-of-custody forms. In this case the name of the carrier is entered under "Received by", the

bill-of-lading number is recorded under "Remarks", and the record is placed in a zipper-lock plastic bag and taped to the inside lid of the shipping cooler.

2. Samples will be packaged properly for shipment and dispatched to the appropriate laboratory via overnight delivery service for analysis, with a separate chain-of-custody record accompanying each shipment (e.g., one for each cooler shipped). Samples are to be shipped within 24 h of collection. Shipping containers will be sealed for shipment to the laboratory, a custody seal will be applied to each cooler to document that the container was properly sealed and to determine if the container was tampered with during shipment.
3. The original chain-of-custody record will accompany the shipment. The copy will be retained by the field operations leader.
4. If sent by overnight mail, the package will be registered with return receipt requested. If sent by common carrier or air freight, proper documentation must be maintained; for example, bill of lading.
5. If samples are split with a source or government agency, a separate chain-of-custody record is prepared for the split samples and marked to indicate the samples are being split. If a representative is unavailable to sign the record, the field operations leader will note it in the field log.

7.3 Laboratory Custody Procedures

Laboratory custody procedures will be equivalent to those described in the latest edition of EPA CLP-IFB Statement of Work. A SOP for laboratory custody procedures is part of the standard laboratory procedures. This QAPjP presents the procedures for ensuring that all data produced follow strict chain of custody procedures.

7.3.1 Sample Receipt

Upon receipt of samples at the laboratory, a designated sample custodian accepts custody of the samples and verifies that the information on the sample labels matches that on the chain-of custody records. The sample custodian will document and notify the supervisor of any discrepancies immediately in the laboratory receiving notebook involving sample integrity, sample breakage, cooler temperature, holding times expiring in transit, appropriate container use, preservatives, and missing or incorrect documentation. The sample custodian will sign and date all appropriate receiving documents.

A sample confirmation letter is sent to the designated project contact within two working days of cooler receipt date or as soon as all the discrepancies have been resolved.

7.3.2 Sample Storage

To assure traceability of samples while in the possession of the laboratory, a method for sample identification that has been documented in a laboratory SOP will be used to assign sample numbers.

Once the samples have been accepted by the laboratory, checked and logged in they are maintained in accordance with laboratory security requirements. Only laboratory personnel are permitted in the sample storage area. All samples must be signed in and out for all analyses. Strict chain of custody procedures are followed.

7.3.3 Sample Analysis

Laboratory personnel are responsible for the custody of the samples until they are returned to the sample custodian.

The following stages of analysis must be documented by the laboratory:

1. Sample extraction/preparation
2. Sample analysis
3. Data reduction and
4. Data reporting

7.3.4 Sample Disposal

When sample analyses and quality assurance checks have been completed in the laboratory, the unused portion of the sample must be stored or disposed of in accordance with EPA CLP protocols. Identifying tags, labels, data sheets, chain-of-custody and laboratory records will be retained until analyses and quality assurance checks are completed in accordance with CLP protocols.

7.3.5 Final Evidence Files

This is the final phase of sample custody. The actual physical sample is stored by the laboratory until the project manager allows its disposal. The chain-of-custody record and sample analysis request form copies held by the laboratory and site manager are archived by both in their respective project files for possible use as evidence in enforcement actions. Laboratory custody forms shall become part of the laboratory final evidence file.

SECTION 8 CALIBRATION AND ANALYTICAL PROCEDURES

This section of the QA Project Plan must describe all of the field and laboratory procedures used for both chemical and physical measurements. Sample preparation methods and cleanup procedures (such as extraction, digestion, column cleaning) should also be included.

All methods must be appropriate for their intended use and be described in sufficient detail. This section, when coupled with QC procedures described in Section 10, should provide enough detail to permit the analytical chemists or other measurement specialists to carry out their procedures unambiguously. This section covers calibration procedures for each analytical or measurement system used to obtain data for critical measurements. Each description should include the specific calibration procedure to be used and frequency of calibration verification.

8.1 Generic Analytical Procedures

Analytical procedures must be established in a data measurement program. These procedures will enable participants in the sampling program to determine the data quality that is to be expected. Each measurement parameter, in the field and laboratory, will follow SOPs to ensure that correct measurements are taken.

The operable unit-specific sampling plans will specify parameters and analyses for sampling activities. Where more than one method is given for a parameter or parameter group, the sampling plan will indicate the appropriate method. In general, methods that provide the most qualitative information will be used during the characterization phases of the project.

Most projects rely heavily on EPA-approved methods that have been validated for environmental samples. Standardized procedures from other organizations, such as the ASTM, and the American Public Health Association, are also commonly employed. When standard methods are unavailable, nonstandard methods may be used but they must be described in detail.

8.1.1 Generic Laboratory Methods

All analytical methods are to be used as written. All changes and modifications to SOPs will be documented in the narrative summary for the data package. All parameters specified by the analytical methods will be determined. Compounds may be added to subsequent analyses if they are identified and judged to be of concern.

All laboratory analyses will be performed by an analytical laboratory with demonstrated proficiency for each parameter.

EPA-approved or similar validated methods can be incorporated by reference, thereby greatly minimizing the effort required to complete this section. Once a method is cited, do not repeat information that is already found in the method. Some additional information is almost always required, however, to assure that project-specific requirements will be met. The following considerations apply:

- It is not necessary to include copies of methods or sections of methods from the Code of Federal Regulations, Standard Methods for the Examination of Water and Wastewater, or Test Methods for Evaluating Solid Waste (SW-846) because these sources are readily available. However, do append ASTM, NIOSH, and IFB procedures because they may not be as readily available to other project principals or reviewers.
- EPA-approved or similarly validated methods that are significantly modified are considered to be invalidated methods and are treated as a nonstandard method.

For projects involving many kinds of measurements, consider providing separate tables for each type of test. Specify the parameter to be measured, the sample type, and the method number (when available).

Certain EPA methods such as those found in SW-846 specify most operating details, including quality control and calibration requirements. Such procedures, however, frequently allow the user to specify certain other options to satisfy project objectives. For example, for multianalyte methods such as GC/MS, the user will typically specify the target compound list that is required for the project. Matrix spike compounds are chosen from the compounds of interest to the particular project, and are not necessarily those recommended in the method. Also list the project-specific target compounds to be used as calibration check compounds and matrix spike compounds. Specify acceptance criteria for matrix spike compounds. In certain cases, isotopically labeled forms of the project analytes may be included as surrogates.

The following is an example of how one might specify various options allowed by Method 8080, a validated method from SW-846 which contains QC and calibration requirements:

> Example: Method 8080.
>
> This method will be employed to determine the 19 pesticides listed in the QA Project Plan. Although PCBs can also be determined by this method, the GC will not be calibrated for PCBs unless they are observed. The matrix spike compounds will be the six critical pesticides listed previously in this QA Project Plan. The surrogates will be tetracholormetaxylene (not dibutylchlorendate). Detection will be by ECD. Quantitation will be by external calibration. Acceptance criteria for surrogate recovery will not be determined by control charts, but must be within a 50 to 150% range. Acceptance criteria for matrix spike/matrix spike duplicates are those stated in Table X of this QA Project Plan. Extraction and cleanup procedures are described below . . .

GC/MS methods may be used to determine the specific method parameters as well as to identify other compounds present that may be of interest. After all compounds of concern are known, other more specific methods may be used wherever available.

- Some EPA-promulgated methods contain only general procedure descriptions and lack specific QC requirements or applicable validation data. For example, EPA Method 18 provides general requirements for GC analysis of stack gases, but contains no QC requirements. For such methods, validation data pertinent to the specific project must be appended. Alternately, a preliminary method validation can be specified as a subtask of the project; in that case, however, specific procedures and acceptance criteria for method validation must also be included as part of the QA Project Plan.
- Other standard methods, such as those found in Standard Methods for the Examination of Water and Wastewater, give operating procedures but omit most QC and calibration requirements. This missing information must be provided either in this section or in Section 8 of the QA Project Plan. As a minimum, be sure to specify the frequency, acceptance criteria, and corrective action plans for all QC procedures and calibrations.

8.1.2 Nonstandard or Modified Methods

Any nonstandard procedure must be described in detail in the form of a Standard Operating Procedure (SOP) that is appended to the QA Project Plan. Validation data applicable to the expected samples must also be included. Validation data must demonstrate that the analytes of interest can be determined without interferences in the expected matrices, and that precision, accuracy, and detection limits will be adequate for the intended use of the data. The method is validated only for samples expected from the specific project, and not for general environmental usage. Once the SOP is written and the validation data are accepted, the method is validated only for use on that specific project and on the specific sample matrix associated with that project. This validation process for a project-specific analysis does not constitute EPA approval for other projects or matrix types.

All nonstandard methods must be validated prior to approval of the QA Project Plan.

The following are minimum guidelines for validating a nonstandard method:

- Determine method detection limits by preparing and analyzing replicates of artificially prepared samples in a matrix similar to the samples of interest. Detection limits should be calculated according to the procedure described in Section 12.1.4 of this document.
- Accuracy should be demonstrated by analyzing spiked samples in the matrix of interest that contains appreciable levels of known or suspected interferences. Accuracy may also be demonstrated by analyzing appropriate Standard Reference Materials (SRMs), if available.
- Demonstrate method precision by replicate analysis on realistic samples.

Each project must be considered on a case-by-case basis, however, and more stringent validation techniques may be required. For example, it might be necessary to perform a comparative analysis with a different analytical technique for some projects.

8.2 Laboratory Analytical Procedures

In accordance with the objectives of the QAPjP, aqueous samples and soil samples will be analyzed according to methods specified in the Scope of Work, EPA-approved procedures, or other validated methods that will meet DQOs requirements. The laboratory will employ analytical methods found in "Test Methods for Evaluating Solid Waste: Chemical/Physical Methods (SW-846), 3rd edition (1986), Office of Solid Waste and Emergency Response". Methods contained in this manual cite the specific calibration and check procedures that are required to conduct the analyses. All appropriate method references must be listed and all methods used must meet the required detection limit.

8.2.1 Organic Methods

This section includes methods for volatiles, semivolatiles, pesticides, PCBs, and other organic analytes. A validated method must exist in the laboratory SOP which includes operation and performance procedures for the GC and GC/MS. For volatiles, the GC/MS is tuned using 50 ng of 4-bromofluorobenzene (BFB). Ion abundance criteria must meet those listed in SW-846. Initial calibration is required at 20 μg/l, 50 μg/l, 150 μg/l and 200 μg/l. For the initial calibration to be considered valid, the relative standard deviation (RSD) must be less than or equal to 30.0% of the response factors for the calibration check compounds. The calibration check compounds are 1,1-dichloroethene, chloroform, 1,2-dichloropropane, toluene, ethylbenzene, and vinyl chloride. Semivolatile calibration procedures follow the same analytical calibration scheme as that of the volatiles with the following differences. Decafluorotriphenylphosphine (DFTPP) is used to meet ion abundance criteria. The initial calibration is required at 20 ng/µl, 50 ng/µl, 80 ng/µl, 120 ng/µl, and 160 ng/µl. There are 13 CCCs and 4 system performance check compounds (SPCC) listed in the method.

GC analyses calibration criteria vary widely depending upon the method quoted. This generally consists of the following. A five-point calibration curve is analyzed and calibration factors are calibrated by either the internal or external standard approach. Percent RSD must then be calculated. Most methods require a percent RSD less than 20%. The calibration is checked on an ongoing basis (generally every ten samples). If the percent difference exceeds that which is required in the method (most methods require 15%), the system is recalibrated.

8.2.2 Inorganic Methods

For metals analyses, two types of analytical methodology are employed: inductively coupled argon plasma emission spectroscopy (ICP) and atomic absorption spectroscopy (AA).

Each ICP is calibrated prior to use using criteria described in the SW-846 protocol. The calibration is verified using standards from an independent source. Interelement correction factors are determined yearly. The linear range of the instrument is established once yearly using a linear range verification check standard. No values are reported above this upper concentration value without dilution. A calibration curve is established daily by analyzing a minimum of two standards, including an initial calibration blank (ICB) and an initial calibration verification (ICV). The ICV must agree within 10% of its true value for the analyses to proceed. The calibration is monitored throughout the run by analyzing a continuing calibration blank (CCB) and a continuing calibration verification standard (CCV) every ten samples. The CCV must agree within 10% of its true value for the data to be deemed acceptable. If this criterion is not met, all samples which are not bracketed by acceptable CCV's must be reanalyzed.

An interelement check standard is analyzed at the beginning and end of each analytical run, to verify that interelement and background correction factors have remained constant. Results outside of established criteria trigger reanalysis of samples.

Each AA unit is calibrated prior to any analyses being conducted. A calibration curve is prepared with a minimum of a calibration blank and three standards and then verified with a standard that has been prepared from an independent source at a concentration near the middle of the calibration range (initial calibration verification — ICV). The ICV must agree within 10% of the true value. The calibration is then verified every ten samples by use of a CCV which must agree within 10% of true value. Results outside of this trigger reanalysis of all samples analyzed since the last acceptable calibration check. All samples for graphite furnace atomic absorption methods are spiked after digestion (analytical spike) to verify the absence of matrix effects or interferences.

Inorganic chemistry methods' calibration and standardization procedures vary depending upon the type of system and analytical methodology required for a specific analysis. For most of the analyses each system is calibrated prior to the analyses conducted. A description of one of the more common calibration approaches is as follows. A five-point curve is generated. A correlation coefficient is determined and must be greater than 0.995. The calibration is checked every ten samples and must agree within 10%, or the ten samples analyzed prior to the unacceptable calibration check are reanalyzed. All requirements are included in the SOPs of the laboratory.

8.2.3 Radiological Methods

For radiological analyses, four types of analytical methodology are employed; gamma spectroscopy, gas proportional counters, alpha spectroscopy, and liquid scintillation counters. All instrumentation must be set up according to manufacturer specification. The laboratory must maintain instrument run log and a maintenance log for each instrument. The specifics of the instruments' initial calibrations and continuing calibrations are included in the laboratory SOPs. The laboratory shall perform calibrations for each radionuclide to be counted, and the standard reference material shall have the same physical form (size, shape, geometry, planchet material, etc.) as the samples to be counted.

The laboratory shall generate self-absorption curves for gross alpha, gross beta, and each individual radionuclide analyses at least every three years using NIST-traceable standards, or equivalent standard reference materials, in which at least 10,000 counts are accumulated. Calibration shall be performed using NIST-traceable standards, or equivalent standard reference material, in which at 10,000 counts are accumulated for 90-Sr, 228-Ra, 137-Cs, and 241-Am.

SECTION 9 DATA REDUCTION, VALIDATION, AND REPORTING

This section describes how data will be reduced, validated and reported. Deliverables that will be required from the analytical laboratory must also be specified. Data generation from sampling activities is usually a four step process: data collection, data reduction, data validation and data reporting. These activities are governed by SOPs and routine reviews which include checking for errors in data entry, data transmission and transcription. This section presents a generic overview of the necessary elements for a generic data reduction, validation, and reporting work plan. This section also describes data reduction, validation and reporting for field investigation activities and laboratory methods.

9.1 Generic Overview: Data Reduction, Validation and Reporting

This section presents the overview of overall data reduction, validation, and reporting principles and typically contains the subsections listed below, plus others at the principal investigator's discretion.

9.1.1 Data Reduction

Name the individuals responsible for data reduction. Summarize the responsibilities of each individual and identify who their managers are. This can be easily represented by an organization chart of all companies involved in the project. Summarize the data reduction procedures that are specific to this project. Data reduction procedures that are part of standard methods or SOPs cited in Section 7 should not be repeated here other than to note any deviations. Summarize the planned statistical approach including formulas, units, and definition of terms. Do not simply reference a "standard text". Example: All results on water samples for radiological testing should be reported in terms of picocuries per milliliters. Explain how results from blanks will be treated in calculations. Example: For all results on water samples for radiological testing, the blank (counts per minute) will be subtracted from sample (counts per minute).

For example, during drilling activities, the field team member supervising a drill rig will keep a chronological log of drilling activities, a vertical descriptive log of lithologies encountered, other pertinent drill information (e.g., staining, odors, field screening, working conditions, water levels, and geotechnical data), and a labor and materials account in the project's bound notebook. After checking the field data and program forms, the data are reduced to tabular form wherever possible, by entering the data into data base files.

9.1.2 Data Validation

Describe the procedures that will be used for determining outliers. Describe the guidelines that will be employed for flagging or validating data. (This requirement is normally satisfied by Section 9 with respect to routine quality control.) In Section 8, discuss criteria that are not covered in Section 9.

Indicate the units for each measurement and each matrix, and whether data will be reported on a wet, dry, or some other reduced basis. If this requirement has been covered in other sections do not repeat it here.

For example, during drilling activities, the field project manager will schedule periodic reviews of archived lithologic samples to ensure that the appropriate lithologic descriptions and codes are being consistently applied by all field personnel.

9.1.3 Data Reporting

Indicate data storage requirements that will be expected of the laboratory once the project is complete. Will the laboratory need to maintain a complete set of raw data for six months? One year? Five years? Also indicate how long the actual samples will be stored, in case reanalysis is required.

List the deliverables that will be expected from the laboratory and from the field operations. Will the data package include all raw data sufficient to recalculate any result, if need be? What type of QC data will be reported? Reporting requirements may vary, depending on the intended use of the data. Summarize the data that will be included in the final report. What QC data will be included? Will analytical and other measurement data be partially reduced before they are reported, or will all individual measurements be reported?

If there are any deliverable requirements above and beyond a standard quote them here.

9.2 RI Field Data Reduction, Validation, and Reporting

This section provides basic concepts of data reduction, validation, and reporting for RI field activities. This section typically contains the subsections listed below, plus others at the principal investigator's discretion.

9.2.1 Introduction

The purpose of this section is to ensure that the data produced by the laboratory are presented in a clear and usable format. In addition, data quality and technical validity must be verified prior to data use. The majority of samples collected at sites will be analyzed according to a specific protocol in which data reduction and reporting schemes are well developed and clearly defined. The employment of these methods ensures comparability with other similarly analyzed environmental samples.

9.2.2 Data Reduction

Data reduction is the process by which raw data generated from the analytical instrument system is converted into usable concentrations. The raw data, which take the form of area counts or instrument responses, are processed by the laboratory and converted into concentrations expressed in terms of parts per million (ppm), parts per billion (ppb), or picocuries per milliliter (pCi/ml). These concentrations are the standard method for expressing the level of contamination present in environmental samples.

The process used to convert the instrument output into usable concentrations is clearly defined in SOPs or statements of work (SOWs) for organic, inorganic and radiological analyses. The SOPs and SOWs present in detail all information, equations, and calculations used. The resulting concentrations are comparable to other environmental samples in general and will be comparable to data previously collected at the sites.

9.2.3 Data Validation

The laboratory is responsible for reviewing data to determine if any analytical problems exist. Specifically, the laboratory will develop a case narrative describing how closely the data meet the DQOs presented in the QAPjP.

Data validation, an after-the-fact review of data, is the process whereby data are determined to be of acceptable or unacceptable quality based on a set of predefined criteria. These criteria depend upon the type(s) of data involved and the purpose for which data are collected. Validation of objective field and technical data will be performed at two different levels. On the first level, data will be validated at the time of collection by following standard procedures and quality control checks. At the second level, data will be validated by the field project manager, who will review the data to ensure that the correct codes and units have been included. After data reduction into tables or arrays, the field project manager will review data sets for anomalous values. Any inconsistencies or anomalies discovered by the field project manager will be resolved immediately, if possible, by seeking clarification from the field personnel responsible for collecting the data.

9.2.4 Data Reporting

The laboratory will report data consistent with reporting requirements. The information supplied in the data package will be sufficient to conduct a thorough validation of the data. The QA reporting for non-CLP data packages will consist of the following accuracy and precision protocols as performed on the appropriate QA samples. For precision, the relative percent difference (RPD) and the percent relative standard deviation (%RSD) will be calculated:

$$\text{RPD} = \frac{d_1 - d_2 \times 100}{(d_1 + d_2)/2}$$

RPD = Relative Percent Difference
d_1 = First Sample Value
d_2 = Second Sample Value (duplicate)

For accuracy, the percent recovery (%R) of spikes will be calculated:

$$\%\text{R} = \frac{\text{SSR} - \text{SR} \times 100}{\text{SA}}$$

SSR = Spiked Sample Result
SR = Sample Result
SA = Amount Spike Added

Field sample precision will be assessed through analysis of duplicate samples and the above RPD equations. Accuracy will be assessed through the analysis of check standards and the above percent recovery equation. Field data will also be assessed in relation to specific project needs. The data package will include the case narrative. This package will include sampling analysis and summary forms.

9.3 Laboratory Data Reduction, Validation, and Reporting

The purpose of this section is to ensure that the data produced by the analytical laboratories are presented in a clear and usable format. In addition, data quality and technical validity must be verified prior to data use. This section will clearly define and develop the data reduction and reporting schemes.

9.3.1 Data Reduction and Validation

All analytical data generated are extensively checked for accuracy and completeness. The data validation process consists of data generation, reduction and review. The analyst who generates the analytical data has the prime responsibility for the correctness and completeness of the data. All data are generated and reduced following protocols specified in the laboratory SOPs. Each analyst reviews the quality of his work based on an established set of guidelines. The analyst reviews the data package to ensure that:

- Sample preparation information is correct and complete
- Analysis information is correct and complete
- The appropriate SOPs have been followed
- Analytical results are complete and correct
- QC samples are within established control limits
- Special sample preparation and analytical requirements have been met
- Documentation is complete (e.g., all anomalies in the preparation and analysis have been documented, holding times are documented, etc.)

The data reduction and validation steps are documented, signed, and dated by the analyst. The analyst then passes the data package to an independent reviewer who performs an independent assessment of the data. This review is conducted according to an established set of guidelines and is structured to ensure that:

- Calibration data are scientifically sound, appropriate to the method, and completely documented
- QC samples are within established guidelines
- Qualitative identification of sample components is correct
- Documentation is complete and correct (e.g., anomalies in the preparation and analysis have been documented, forms are complete, holding times are documented, etc.)
- The data are ready for incorporation into the final report
- The data package is complete and ready for data archiving

This review is structured so that all calibration data and QC sample results are reviewed and all of the analytical results from 10% of the samples are checked back to the benchsheet. If no problems are found with the data package, the review is complete. If any problems are found with the data package, an additional 10% of the samples are checked to the benchsheet. The process continues until no errors are found or until the data package has been reviewed in its entirety. The signature and date by the reviewer indicates the final report can be prepared and approved.

Before the report is released to the client, the program administrator reviews the report to ensure that the data meet the overall objectives of the client.

9.3.2 Data Reporting

In general, reports contain the following items.

- **General discussion** — Descriptions of sample types, tests performed, problems encountered, and general comments are given.
- **Analytical data** — Data are reported by sample by test and are not blank-corrected. Pertinent information including dates sampled, received, prepared, and extracted are included on each result page. The reporting limit for each analyte is also given.
- **QC information** — Analytical results for laboratory blanks are reported where applicable. In addition, the results (average percent recovery and relative percent difference) of the DCS analyzed with the project are listed. Control limits are reported.
- **Methodology** — References for analytical methodologies used are cited.

9.3.3 Project Files

Project files are created for each project handled within the laboratory. These files contain all documents associated with the project. This includes correspondence from the client, chain-of custody records, raw data, copies of laboratory notebook entries pertaining to the project, and a copy of the final report. When a project is complete, all records are passed to the document custodian who puts the files into the document archive. All files are secured in limited access areas and are signed in and out of the area under chain of custody. Raw data and all pertinent records are retained for a minimum of [laboratory fill in] years.

SECTION 10 INTERNAL QUALITY CONTROL CHECKS

This section describes all internal quality control (QC) checks that will be used throughout the project, including field and laboratory activities of all organizations involved. The QC procedures that are specified should follow from the QA objectives stated in Section 3. Thus, Section 3 specifies the analytical requirements, while Section 10 describes how these specifications will be met.

This section of the QAPjP must provide an unambiguous description of QC procedures. There is no need, however, to repeat in the test any material that is already described in a standard method.

10.1 Types of QC Checks

Examples of QC checks that should be considered include the following:

- **Samples** — Co-located, split, replicate
- **Spikes** — Matrix spikes and matrix spike duplicates; spiked blanks; and surrogates and internal standards
- **Blanks** — Sampling, field, trip, method, reagent, instrument and zero and span gases
- **Others** — Standard reference materials (complex natural materials, pure solutions); mass tuning for mass analysis; confirmation on second column for gas chromatographic analyses; control charts; independent check standard; determinations of detection limits; calibration standards; proficiency testing of analysts; any additional checks required by the special needs of your project

Include QC checks for process measurements as well.

Be sure to: Identify the stage at which replication and spiking occur. Avoid using terms such as "sample replicate" without explaining how such replication will be performed. Explain exactly how blanks will be prepared.

10.2 Items To Include

Most information for this section can be summarized in a table. This table should designate the types of QC procedures, required frequencies, associated acceptance criteria, and corrective action that will occur if the acceptance criteria are not met. When QC procedures are referenced to a standard method that describes an exact procedure, additional discussion is not normally needed. However, standard methods lacking exact QC procedures or nonstandard methods require a detailed explanation, either in this section of the QA Project Plan or in an appended Standard Operating Procedure.

The tabular format is also convenient for summarizing routine and ongoing calibration requirements. If routine calibration is summarized in this section, a reference to that effect should be included in Section 7.

The accompanying text must assure that there are no ambiguities. Particularly troublesome terms are "duplicate" or "replicate". The text should explain precisely how and when replicates are taken. Do replicate samples, for instance, refer to samples collected simultaneously or sequentially in the field; to samples collected at the same sample point but at different times; to samples that are split upon receipt in the laboratory? The term "QC check sample" must also be carefully defined. Indicate at which point matrix spiking occurs.

Exact procedures for preparing the numerous kinds of blanks must be described fully in the text. Never assume that a term such as "field blank" will mean the same to the sampling team or to a reviewer that it does to you.

Either in this section or in Section 7, be sure to specify what compounds or elements will be employed as matrix spikes and surrogates. Some standard methods recommend surrogate and matrix spike compounds, which may be incorporated by reference when appropriate. Typically for some projects, some of the matrix spike compounds must be selected on a project-specific basis.

In some cases, it may also be necessary to provide additional discussion of potential problems that might be expected with certain QC procedures, along with the proposed solutions. For example, spiking samples in the field is frequently less reliable and more difficult than spiking in the laboratory, due to contamination and less controlled conditions. If field spiking is required, then a discussion of procedures that minimize such problems is also required.

Because standard methods often include extensive QC requirements, why must such QC procedures be summarized in this section? Why not simply state that QC will be performed as required in the method? In some cases, standard methods are sufficiently complete for this approach, but frequently they are not. First, many methods do not include specific QC procedures. Second, even the more complete methods allow options, such as the choice of matrix spike compounds, or the use of either control charts or fixed acceptance limits. Third, the analytical and measurement requirements are project-specific and require project-specific QC guidelines.

10.2 RI Field Quality Control

In order to monitor the quality of analytical data generated for a RI, an appropriate number of quality control (QC) methods will be employed for all field measurement systems. Employing QC methods permits the validation of the analytical methodology utilized and provides a measure of the suitability of the methodology to meet the DQOs prior to the beginning of measurements or analysis. Once measurement and analysis has begun, employing QC methods permits the monitoring of the system output for quality. The QC results, presented with the environmental sample data, allows the data to be assessed for quality, and a determination made on how well the data have met the DQOs. Field-generated data are used in conjunction with laboratory data for further investigation of contamination at the site. Both laboratory and field internal QC programs include steps to assure the data are reliable for the extent they will be used in the RI.

10.2.1 Field Quality Control Measurements

The intended data uses have been identified and the DQOs established for all field measurement activities in Sections 3 and 5 of this QAPP. The required field measurements and instruments to be used are given in the FSP of the SAP. The FSP contains SOPs which describe the use and calibration of field instruments. QC methods will be used to demonstrate that the instruments are capable of producing reliable data. The calibration check samples will be analyzed daily and duplicate samples will be analyzed at a minimum frequency of 5%. The calibration check verifies that the instrument is capable of accurately identifying and quantifying contaminants of concern. The duplicates provide a quantitative measurement of the precision of the instrument. Background samples are similar to blanks and provide information regarding instrument reliability. The information is recorded in the field log books. The results from these QC methods are used by field technicians to monitor the instrument at the time of the analysis. If QC results indicate a problem with the instrument, corrective action will be taken and, if necessary, the samples will be reanalyzed. Because field measurements are generally easy to repeat, measurements should be repeated as necessary so the data are as complete as possible. The QC results are used as an indication of data quality and reliability when the data are being reviewed.

10.2.2 Soil Sampling

Quality control checks for field soil sampling activities will follow guidance provided in the Soil Sampling Quality Assurance User's Guide (EPA 1984b). Field replicate soil samples will be given a unique alphanumeric identifier and submitted to the laboratory blind (i.e., without indicating the location). These samples will serve as blind field splits and will be used to evaluate laboratory reproducibility and field reproducibility. Field duplicate soil samples will be identified on the appropriate forms. Trip blanks and equipment (rinsate) blanks are not suggested Quality Assurance/Quality Control (QA/QC) procedures for soil samples according to the U.S. EPA Soil Sampling Quality Assurance User's Guide (EPA 1984b). Therefore, trip blanks will not be part of QA/QC programs for soil sampling Activities. Equipment blanks, however, will be included as part of the field QA/QC program for soil sampling activities. Since radiological contamination may be encountered at any locations, the addition of equipment blanks will serve as a check on the sampling device cleanliness.

10.2.3 Water Sampling

Field duplicate water samples will be given a unique alphanumeric identifier and submitted to the laboratory blind (i.e., without indicating the location). These samples will serve as blind field splits and will be used to evaluate laboratory reproducibility and field reproducibility. Field duplicate water samples will be identified on the appropriate forms.

10.3 Laboratory Quality Control

A summary of laboratory-based quality control samples and frequency is presented in SW-846 methods. At a minimum, one laboratory matrix spike and replicate per 20 environmental samples or analytical batch, whichever is more frequent, will be used. Reagent blanks will be at a frequency of one per analytical batch. Three types of laboratory standards will be used: (1) calibration standard, (2) check standards, and (3) quality control reference samples. Calibration standards are prepared by diluting the stock analyte solution in graduated amounts which cover the expected range of the samples being analyzed. Calibration standards must be prepared using the same type of acid or solvent as will result in the samples following sample preparation. These criteria are applicable to organic, inorganic, and radiological analyses. Results obtained from analysis of standards are used to generate a standard curve which plots concentrations of known analyte standards versus the instrument response to the analyte. The standard curve is used to quantitate the compound in an environmental sample. A minimum of three calibration standards will be used to generate a standard curve for all inorganic analyses and a five-point minimum calibration curve will be used for organic analyses. A check standard is a material of known composition that is analyzed concurrently with test samples to evaluate a measurement process. The check standard is prepared by the analyst to monitor and verify instrument performance on a daily basis. A quality control reference sample is prepared from an independent standard at a concentration other than that used for calibration, but within the calibration range. An independent standard is defined as a standard composed of the analyte(s) of interest from a different source than that used in the preparation of standards for the standard curve. The quality control reference sample serves as an independent check for technique, methodology, and standards.

10.3.1 Laboratory QC Checks

Laboratory QC protocols for analytical analyses include the following items:

- A minimum of one method blank is analyzed per sample batch to detect contamination during preparation and/or analysis.
- Duplicate control samples (DCS) consisting of target analytes spiked into a blank matrix and analyzed for every 20 samples to determine accuracy and precision.
- Matrix spikes and matrix spike duplicates for organic and inorganic analyses will be analyzed for every 20 samples to determine the affect of the matrix on the method performed. It is the client responsibility to collect sufficient sample for project-specific QC.
- Internal and surrogate standards will be added where appropriate to quantitate results, determine recoveries, and to account for sample to sample variation.
- Calibration of instrumentation will be determined according to the appropriate DOE approved methods.

10.3.2 Specific QC Assignments by Sample Group

Specific laboratory QC samples which will be analyzed per sample group are as follows:

- **Organic** — DCS: Per 20 samples; MS/DU: Per 20 samples per matrix; and MB: Per sample batch per matrix
- **Inorganic** — DCS: Per 20 samples; MS/DU: Per 20 samples per matrix; and MB: Per sample batch per matrix where appropriate

Note:

1. It is the responsibility of the client to collect sufficient sample and designate MS/SD/MD analyses on the chain of custody
2. Accuracy and precision are determined through the results of the DCS
3. DCS = duplicate control sample; MS = matrix spike; SD = matrix spike duplicate; DU = matrix duplicate; and MB = method blank

SECTION 11 PERFORMANCE AND SYSTEM AUDITS

11.1 Generic Performance and System Audits

Each QA Project Plan should describe the QA audits planned for monitoring the system to be used for obtaining critical measurements. A schedule of all contractor-planned audits should be included, along with identification of the responsible personnel. This section should also indicate what audit reports will be generated and who will receive these reports. If no audits are planned, include an explanation.

QA audits may include one or more Technical Systems Audits (TSAs), Performance Evaluation Audits (PEAs), and Audits of Data Quality (ADQs).

A TSA is a qualitative audit of all components of the total measurement system, including technical personnel and QA management. This type of audit includes a careful evaluation of both field and laboratory QC procedures. TSAs are normally performed before, or shortly after, measurement systems are operational; they should also be performed on a regularly scheduled basis throughout the lifetime of the project.

After measurement systems are operational and begin generating data, PEAs are conducted periodically to determine the bias of the total measurement system(s) or component parts. As part of a PEA, the laboratory analyzes a performance evaluation sample. QA Project Plans should also indicate any scheduled participation in other interlaboratory performance evaluation studies. Long-term projects should provide for regularly scheduled PEAs.

ADQs are retrospective evaluations of data. Typically, a representative portion of the results in an analytical report is reviewed in detail, starting with raw data and chromatograms, and proceeding through the calculation of final results. ADQs are often used to resolve specific questions regarding the quality of a data set.

11.2 RI Field Quality Assurance Audits

11.2.1 Introduction

In order to monitor the capability and performance of all RI activities, audits will be conducted by QA personnel. Technical systems audits (TSAs) are conducted to determine the suitability and capability of project activities to meeting project quality goals. TSAs include on-site field audits to monitor the field techniques, procedures, and the overall implementation of the QAPP procedures. These will be conducted periodically by the site quality assurance officer (QAO). Performance audits (PAs) are conducted to measure the accuracy of operating measurement systems. Data quality audits (DQAs), are conducted to determine if the data generated by the sampling and analysis satisfy the predetermined DQOs. The site QAO will be responsible for conducting DQAs of all data generated from project activities. The site QAO will be responsible for conducting TSAs and PAs.

The PMO QA manager will perform periodic audits of the project QAO efforts to review whether QA objectives are being met. In addition to this audit, one program audit will be conducted per year by the Corporate QAO and the PMO QA Manager, to assure that program QA objectives are being met at the project QA level.

11.2.2 Technical Systems Audits (TSAs)

Technical systems audits (TSAs) consist of an evaluation and review of all components of a measurement system to determine its capability to meet project quality goals, and to determine if the procedures of the system are being properly followed.

The following components of each measurement system will be reviewed, with other items added as necessary:

1. Sample collection and analytical activities
2. Equipment calibration techniques and records
3. Decontamination and equipment cleaning
4. Equipment suitability and maintenance/repair
5. Background and training of personnel
6. Sample containers, preservation techniques, and chain of custody
7. Data log books
8. Monitoring well installation and development

TSAs are conducted prior to the operation of each measurement system to determine if the system is capable of producing data that will meet the DQOs. This initial audit includes a careful evaluation of both field and laboratory QC procedures. Once the system is approved and shortly after it becomes operational, a TSA is conducted to monitor the performance of the measurement activities. This includes an audit of field procedures to determine if the appropriate SOPs are being followed. In addition, TSAs will determine if this QAPP is being implemented in the field.

Following the initial audits, TSAs will be conducted on a regularly scheduled basis. A written QA audit report will be prepared by the site QAO and submitted to the site manager and the PMO QA manager. The report will identify any deficiencies found and recommend corrective action. The PMO QA manager will assist with corrective action and maintain a log of the audit activities. Follow-up reports describing corrective actions which have been completed will be submitted by the site manager to the PMO QA manager.

11.2.3 Performance Audits (PAs)

A performance audit (PA) is conducted on all laboratories through analysis of a performance evaluation (PE) sample. The PE sample is a sample of known analyte concentration that is analyzed by the laboratory. The analytical results detected by the laboratory are compared with the known results. The results provide a measure of laboratory performance that is used along with other QA criteria to monitor laboratory capability. The EPA administers required PAs to CLP laboratories every six months. PAs of non-CLP laboratories will be conducted by staff personnel.

11.2.4 Data Quality Audits

Data quality audits (DQAs) are conducted to determine if the data are adequate to support the DQOs and to determine the cause of deficiencies in the event that the data quality is not adequate. This audit will be conducted by the site QAO after the data have been fully validated. The Site QAO will first determine to what extent the data can be used to support the decision-making process. Secondly, the site QAO will identify the cause of any deficiencies in the data, whether technical, managerial, or both. Finally, the site QAO will submit a written DQA report to the site manager and the PMO QA manager.

11.3 Laboratory Performance and System Audits

11.3.1 External Audit by Regulatory Agencies: DOE/EPA/Others

The AAA Laboratory participates in a wide variety of certifications, programs, and contracts and is subjected to rigorous external audits by many government agencies and private clients.

The AAA Laboratory is a CLP laboratory and presently holds an EPA CLP contract for organics, and is audited on a regular basis by the U.S. EPA under this contract. Quarterly performance evaluations are also performed under this contract in addition to participating in U.S. EPA/WP series performance evaluation samples. (Address all other certification programs that the lab participates in.)

11.3.2 Internal Audits

The laboratory is subjected to periodic systems audits by the QA department. These audits are intended to serve two purposes:

1. To ensure that laboratory is complying with the procedures defined in laboratory SOPs, QAPjPs, and contracts.
2. To determine any sample flow or analytical problems. The frequency of the audits will be increased if any problems are suspected.

A corporate QA audit is performed on an annual basis by the corporate director of quality assurance. This audit is intended to check compliance with the AAA Laboratory overall QA program.

All audits by divisional and corporate QA staff are performed more frequently, or specifically directed audits are performed if any problems are suspected in the laboratory.

SECTION 12 PREVENTIVE MAINTENANCE

12.1 Generic Preventive Maintenance

For most projects, instrument performance is checked immediately before and after use, and in such instances preventive maintenance is not an issue. On long-term monitoring programs, preventive maintenance can become an important tool, especially assuring that QA requirements are met. All remediation projects must provide a brief summary of preventive maintenance, or should state why it is not relevant to the project.

This section should include a summary description of preventive maintenance procedures, a schedule for performing these procedures, a list of major spare parts, and lists of maintenance contracts for critical measurement systems. For convenience, these items may be presented in tabular format.

12.2 RI Field Preventive Maintenance

12.2.1 Purpose

The purpose of the preventive maintenance program is to ensure that the sampling, field testing and analytical equipment perform properly, thereby avoiding erroneous results, and minimizing equipment downtime. The preventive maintenance program also provides for the documentation of all maintenance to be used as evidence of instrument maintenance and for scheduling of future maintenance. This section describes the equipment maintenance program for field instruments and those responsible for implementation of the program at the XYZ Landfill site. The specific equipment maintenance procedures are given in the equipment SOPs and the preventive maintenance SOPs. The laboratory preventive maintenance program is the responsibility of the laboratory and only the minimum requirements are mentioned here.

12.2.2 Responsibilities

Title	Responsibilities
Field operation leader	Keeping all maintenance records. Development and implementation of maintenance program.
Equipment manager	Maintaining storage of equipment within the company inventory. Carrying out all maintenance according to schedule. Informing field team members of specific maintenance requirements.
	Keeping records of all maintenance performed under his care. Sending out equipment for service/repair. Maintaining adequate supply of spare parts.
Field personnel	Maintenance of all equipment located on-site on a regular basis and after each use. Keeping supply of spare parts on-hand.

12.2.3 Field Equipment

Field equipment will be properly calibrated, charged, and in good general working condition before the beginning of each working day. The program SOPs define the required equipment checks and calibration requirements for each type of field equipment. Field equipment which does not meet the calibration requirements will be taken out of service until acceptable performance can be verified. Nonoperational field equipment will also be removed from service and returned to the supplier, and a replacement will be obtained. Maintenance records will be maintained for each field instrument according to a unique number affixed to the instrument. These records will be reviewed prior to their use in the field to assure that instrument maintenance and calibration are up to date.

All field instruments will be properly protected against inclement weather conditions during environmental investigations. At the end of each working day, all field equipment (except self-propelled equipment) will be taken out of the field and placed in appropriate storage.

All self-propelled field equipment (e.g., drill rigs, water trucks, and support vehicles) will arrive at the site in proper working condition each day. All lubricating, hydraulic, and motor oils will be checked before the start of each work day to make certain all fluid reservoirs are full and there are no leaks. If a leak is detected, the equipment will be removed from service for repair or replacement.

12.2.4 Preventive Maintenance Program

The preventive maintenance program consists of three parts, normal upkeep, service and repair, and formal record keeping. Normal upkeep consists of daily procedures that include cleaning, lubrication, and checking the batteries of the equipment. The following is a partial list of normal upkeep procedures and a partial list of important spare parts:

Normal upkeep for environmental monitoring equipment performed daily or after each use

1. Cleaning (see instrument SOPs)
2. Lubrication of moving parts
3. Check/charge battery
4. Inspect for damage
5. Check for operation problems
6. Inspect all hoses and lines

Partial list of important spare parts for environmental monitoring instruments planned for use at the AAA Landfill site:

1. Fuses
2. HNU-UV lamp
3. Probes
4. Spare battery
5. Gas refills
6. Septa
7. Strip chart paper
8. Extra syringes

9. Teflon tape
10. OVA — Igniter; filter cup; and particle filters

The normal upkeep is performed daily after each use and includes inspecting for damage, signs of problems, and charging the batteries if necessary. Specific equipment upkeep procedures are described in the SOP.

Minor service and repair will be performed by the equipment manager who is trained in the service and repair of field instruments. Equipment in need of major or more complex repair and service will be sent to the manufacturer.

All maintenance, servicing, and repair of equipment shall be recorded and kept on file. Field personnel shall record maintenance and instrument problems in the field instrument log books. These will ultimately be kept on file by the field operations leader. The equipment manager shall keep a record of all equipment released to the field and a record of all maintenance and service on file.

12.2.5 Rental Equipment

Rental equipment will be obtained from a rental supplier. The equipment will require a pre-receipt to verify accuracy, maintenance and upkeep of the equipment.

12.3 Laboratory Preventive Maintenance

Preventive maintenance procedures will be clearly defined and written for each measurement system. Maintenance activity, preventive or repair, will be documented on standard forms which are maintained in log books. Written procedures will include maintenance schedules, problem identification procedures, space for describing problems and repair notes, and failure analysis protocols. Service contracts and regularly scheduled in-house maintenance will be included, along with a list of critical spare parts.

12.3.1 Laboratory Equipment

The ability to generate valid data requires that all analytical instrumentation be properly maintained. Service contracts will be provided for all major equipment to provide routine preventive maintenance and emergency repair service. The elements of preventive maintenance programs for each piece of equipment are detailed in their respective SOP.

12.3.2 Instrument Maintenance Logbooks

The laboratory personnel will document all instrument maintenance and repair in the specific instrument maintenance logbooks.

12.3.3 Instrument Calibration and Maintenance

Major analytical instruments have specific preventive maintenance and calibration schedules. Preventive maintenance and calibration is generally performed by the manufacturer's service representatives on a routine basis.

12.3.4 Spare Parts

The laboratory will maintain an inventory of routinely required spare parts relevant to the services provided (i.e., sources, vacuum pumps, and filaments for GC/MS; torches and burner heads for AA/ICP).

SECTION 13 PROCEDURES TO ASSESS DATA QUALITY INDICATORS

This section describes how data quality indicators will be calculated and reported. As a minimum, equations must be provided for precision, accuracy, completeness, and method detection limits. In addition, equations must be given for other project-specific calculations, such as mass balance, emission rates, confidence ranges, etc.

Make sure that this section complements Section 5 (QA Objectives) and Section 10 (Internal QC Checks) of the QA Project Plan. Section 5 specifies which particular data quality indicators will be employed. Section 10 then uses these specific indicators to generate acceptance criteria. Because groups can use different equations to calculate data quality indicators, Section 13 must provide the exact equations to avoid misunderstandings among future data users.

13.1 Generic Calculation of Data Quality Indicators

Listed below are general guidelines for calculating the more common data quality indicators.

13.1.1 Common Data Quality Indicators

Precision

If calculated from duplicate measurements, relative percent difference is the normal measure of precision:

$$\mathrm{RPD} = \frac{(C_1 - C_2) \times 100\%}{(C_1 - C_2)/2} \qquad \mathbf{(1)}$$

where: RPD = relative percent difference; C^1 = larger of the two observed values; and C^2 = smaller of the two observed values.

If calculated from three or more replicates, use relative standard deviation rather than RPD:

$$\mathrm{RSD} = (s/\bar{y}) \times 100\% \qquad \mathbf{(2)}$$

where: RSD = relative standard deviation; s = standard deviation; and $\bar{y}$ = mean of replicate analyses.

Standard Deviation

$$s = \sqrt{\sum_{i=1}^{n} \frac{(y_i - \bar{y})^2}{n-1}} \quad \textbf{(3)}$$

where: s = standard deviation; y_i = measured value of the ith replicate; $\bar{y}$ = mean of replicate measurements; and n = number of replicates.

For measurements, such as pH, where the absolute variation is more appropriate, precision is usually reported as the absolute range, D, of duplicate measurements:

$$D = |m_1 - m_s| \quad \textbf{(4)}$$

where D = absolute range; m_1 = first measurement; and m_2 = second measurement.

The standard deviation, s, given above, can also be used.

Accuracy

For measurements where matrix spikes are used, calculate the percent recovery as follows:

$$\%R = 100\% \times \left(\frac{S - U}{C_{sa}} \right) \quad \textbf{(5)}$$

where $\%R$ = percent recovery; S = measured concentration in spiked aliquot; U = measured concentration in unspiked aliquot; and C_{sa} = actual concentration of spike added.

When a standard reference material (SRM) is used:

$$\%R = 100\% \times \left(\frac{C_m}{C_{srm}} \right) \quad \textbf{(6)}$$

where $\%R$ = percent recovery; C_m = measured concentration of SRM; and C_{srm} = actual concentration of SRM.

Completeness

$$\%C = 100\% \times \left(\frac{V}{n} \right) \quad \textbf{(7)}$$

where $\%C$ = percent completeness; V = number of measurements judged valid; and n = total number of measurements necessary to achieve a specified level of confidence in decision making.

Note: This more rigorous definition of completeness is an improvement on the conventional definition in which "*n*" is replaced by "T", the total number of measurements.

Method Detection Limit (MDL)

MDL is defined as follows for all measurements:

$$MDL = t_{(n-1,\ 1-\alpha=0.99)} \times s \quad \textbf{(8)}$$

where MDL = method detection limit; s = standard deviation of the replicate analyses; and $t_{(n-1,\ 1-\alpha\ =\ 0.99)}$ = student T-value for a one-sided 99% confidence level and a standard deviation estimate with n–1 degrees of freedom.

13.1.2 Project-Specific Indicators

Projects frequently incorporate data quality indicators in addition to those discussed in the previous sections. The following is an example of a project-specific data quality indicator:

Example: Mass balance calculation for a soil washing process being tested with PCP-contaminated soils. Mass balance (MB) will be calculated according to

$$MB = M_{out}/M_{in} \quad \textbf{(9)}$$

where M_{out} and M_{in} denote the total mass of PCP in the output and input streams for each test condition.

13.2 RI Field Data Assessment

13.2.1 Overview

All analytical data received from the analytical laboratories will be assessed to determine to what extent the data can be used in making sound project decisions. The goal of data assessment is to characterize the data so that project decisions are made using data that is of sufficient quality to support those decisions. The levels of quality needed to support the various project decisions have been stated in the form of the DQOs. Where the DQOs are met, the data are useful in making necessary decisions.

In order to determine how well the DQOs have been met, all CLP and non-CLP data will be reviewed and validated by project personnel. The data will be reviewed and validated with the intended data uses and DQOs being utilized to aid in decisions regarding data usefulness.

13.2.2 Data Assessment

Organics

Organic data will be assessed for the following QA parameters:

- Sample holding times
- GC and GC/MS tuning and performance
- Calibration
- Blanks
- Surrogate recovery
- Matrix spike/matrix spike duplicate
- Compound identification
- Internal standards
- System performance
- Overall assessment of data

Inorganics

Inorganic data will be assessed for the following QA parameters:

- Holding times
- Blanks
- Calibration checks
- ICP interferences
- Duplicates
- Matrix spikes
- Field QC
- Package completeness

13.2.3 Reports

Data review and validation reports shall be generated to describe the validation and any problems encountered, as well as to provide data summaries in which the data are appropriately qualified. These reports will be used in applying the data as part of the RI studies.

13.3 Laboratory Calculation of Data Quality Indicators

By following all the procedures outlined in this QAPjP and by thoroughly documenting all work that is performed, Enseco will closely monitor data precision, accuracy, and completeness. Validity of reporting limits is also assured.

13.3.1 Data Quality

For this project, the methods to determine precision and accuracy and their acceptability are well defined in the data quality objectives section and in the analytical methods.

13.3.2 Precision

Precision is determined by the comparison of duplicate control samples. The RPD of duplicate control samples will be used to estimate the precision. The following equation will be used to determine this.

$$\text{RPD} = \frac{|d_1 - d_2|}{(d_1 + d_2)/2} \times 100$$

where RPD = relative percent difference; d_1 = first sample value; and d_2 = second sample value (duplicate).

13.3.3 Accuracy

The determination of accuracy of a measurement requires a knowledge of the true or accepted value for the analyte being measured. The average percent recovery of duplicate control samples will be used to estimate accuracy. Accuracy will be calculated in terms of average percent recovery in the following equation.

$$\text{Average percent recovery} = 100 \times \frac{\overline{X}}{T}$$

where $\overline{X}$ = average of observed value(s) for measurement(s); and T = "true" value.

13.3.4 Analytical Completeness

Determining whether a data base is complete or incomplete is a subjective evaluation. to be considered complete, the data set must contain all QC check analyses verifying precision and accuracy for all of the analytical protocols. Less obvious is whether that data are sufficient to achieve the goals of the project. All data are reviewed in terms of goals in order to determine if the data base is sufficient.

Percent completeness is calculated as follows:

$$\text{Completeness} = \frac{\text{valid data obtained}}{\text{total data obtained/collected}} \times 100$$

13.3.5 Detection Limits

The sensitivity of an analytical method is related to the detection limit (i.e., the lowest concentration of an analyte that can be detected at a specific confidence level). Definitions of instrument detection limit (IDL), method detection limit (MDL), and practical quantitation limit (PQL) follow in this section.

- **IDL** — This is the smallest signal above background noise that an instrument can detect at a 99% confidence level. An IDL is measured by analyzing three replicate standards. It is calculated as three times the standard deviation of the replicate analyses. IDLs are determined for metals analyses.
- **MLD** — This is the minimum signal level required to qualitatively identify a specific analyte by a specific procedure at a greater than 99% confidence interval. An MDL is measured by analyzing seven replicates spiked at one to five times the expected method detection limit. It is calculated by multiplying the standard deviation times the student *T*-value at the desired confidence level. The laboratory uses a 99% confidence interval and

seven spiked replicates of a control matrix in determination of method detection limits. This is performed as referenced in 40 CFR Part 136, Appendix B.
- **PQL** — This is the minimum level that can be reliably achieved by a method within specified limits of precision and accuracy. The laboratory's PQL is derived from the evaluation of interlaboratory method detection limit studies. This is the reporting limit.

The laboratory determines the MDL for routine organic methods and IDLs for metals using a blank matrix. These data are kept on file in the QA office.

SECTION 14 CORRECTIVE ACTION

Each QA Project Plan must incorporate a corrective action plan. This corrective action plan must include the predetermined acceptance limits, the corrective action to be initiated whenever such acceptance criteria are not met, and the names of the individuals responsible for implementing the plan.

14.1 Generic Corrective Action

Routine QC procedures already included in Section 10 need not be repeated here. This section is reserved primarily for nonroutine corrective action not described elsewhere. Nonroutine corrective action may result from common monitoring activities, such as:

- Performance evaluation audits
- Technical systems audits
- Interlaboratory comparison studies

It may also arise from conditions unique to specific projects, as seen below.

Example: Bio-oxidation study of hazardous waste in municipal sludge carried out over a one-year period.

The prime contractor's quality assurance manager will submit to the laboratory a blind performance evaluation (PE) sample containing some or all of the target analytes before any analytical work begins and on a monthly basis thereafter. The average percent recovery of all target analytes must be between 80 and 120%, with no outliers less than 50 or greater than 150%. If these limits are exceeded, analytical work will stop until the problems are identified and solved. Before work is restarted another blind PE sample must be analyzed and results must meet the acceptance criteria. Results of these PE samples will be included in the final report.

14.2 RI Field Corrective Action

14.2.1 NonConformance Reports

Corrective action will be undertaken when a nonconforming condition is identified. A non-conforming condition occurs when QA objectives for precision, accuracy, completeness, representativeness, or comparability are not met, or when procedural practices or other conditions are not acceptable.

A nonconformance report will be prepared by the site QA officer, approved by the PMO QA manager, and issued to the site manager and other appropriate parties as described below in Section 14.2. The nonconformance report will describe the unacceptable condition and the nature of corrective measures recommended. A schedule for compliance will also be provided.

14.2.2 Corrective Action

The nonconformance report will be transmitted to the program manager, PMO QA manager, and the site manager. The nonconformance report will specify, in writing, the corrective action recommended, including measures to prevent a recurrence of the original deficiency. Appropriate documentation of corrective action will also be prepared. The site QA officer will monitor implementation of the correction action, and provide written record as to whether the original problem has been resolved.

14.2.3 Stop-Work Order

A stop-work order may be issued, upon authorization, by the site QA officer, if corrective action does not adequately address a problem or if no resolution can be reached. To issue a stop-work order, written authorization is required from the site manager. If disagreement occurs among these individuals, it will be brought before successively higher levels of management until the issue is resolved.

14.2.4 Documentation of the Stop-Work Order

The condition and need for a stop-work order will be documented in sufficient detail to permit evaluation of the deficiency and determination of proper corrective action. Pertinent communications will be attached to the stop-work order and referenced in the appropriate spaces. Such communications include discussions, correspondences, or telephone conversations which pertain to evaluation of the problem and potential solutions, and implementation of the preferred solution.

14.2.5 Resumption of Work

In order for work to resume following a stop-work order, the site manager must rescind it in writing.

14.2.6 Course and Action To Prevent Recurrence

The site QA officer is responsible for tracking nonconforming conditions, evaluating the effectiveness of corrective measures, and assuring that the necessary steps have been taken to prevent recurrence of the original problem.

14.2.7 Field Changes

The site manager is responsible for all site activities. In this capacity the site manager will at times be required to modify site programs in response to changing site conditions. At such times the responsible field operations leader will notify the site manager of the anticipated change, and obtain the approval of the site manager and implement the necessary changes. The site manager will notify in writing the site QA officer, and the operations manager. A copy of the notification will be attached to the file copy of the affected document. If an unapproved action has been taken during a period of deviation, the action will be evaluated to determine the significance of any departure from established procedures.

Changes in the program will be documented on a field change request which is signed by the field operations leader and the site manager. The site manager will maintain a log for the control of field change requests.

The site manager is responsible for controlling, tracking, and implementing the identified changes. Completed field change requests are distributed to affected parties which will include as a minimum: PMO operations manager, site manager, site QA officer, and field operations leader.

14.3 Laboratory Corrective Action

Corrective actions for laboratory problems are specified in laboratory SOPs. Specific QC procedures are designed to help analysts determine the need for corrective action. Often, personal experience is most valuable in alerting the analyst to suspicious data or malfunctioning equipment. Corrective action taken at this point helps to avoid collection of poor-quality data.

Problems not immediately detected during the course of analysis may require more formalized, long-term corrective action. The essential steps in the corrective action systems are as follows:

1. Identify and define the problem
2. Assign responsibility for investigating the problem
3. Investigate and determine the cause of the problem
4. Determine a corrective action to eliminate the problem
5. Assign and accept responsibility for implementing the corrective action
6. Establish effectiveness of the corrective action and implement it
7. Verify that the corrective action has eliminated the problem

This scheme is generally accomplished through request to the QA department. Any laboratory analyst or project member may notify the QA team of a problem. The QA team initiates the corrective action scheme by relating the problem to the appropriate team managers and/or program administrators who investigate or assign responsibility for investigating the problem and its cause. Once determined, an appropriate corrective action is approved by the QA team. Its implementation is later verified through an audit.

SECTION 15 QUALITY ASSURANCE REPORTS TO MANAGEMENT

15.1 Generic Quality Assurance Reports To Management

This section of a QA Project Plan identifies the individuals responsible for QA reports, and describes the type and frequency of reports (weekly oral presentations and discussions, monthly written reports, etc.) that will be used to keep project management informed. As a minimum, such reports include:

- Changes in the QA Project Plan
- Summary of QA/QC programs, training, and accomplishments
- Results of technical systems and performance evaluation audits
- Significant QA/QC programs, recommended solutions, and results of corrective actions
- Data quality assessment in terms of precision, accuracy, representativeness, completeness, comparability, and method detection limit
- Discussion of whether the QA objectives were met, and the resulting impact on decision making
- Limitations on use of the measurement data

Managers receiving these detailed reports will then be able to monitor data quality easily and effectively.

If subcontractors are used, include a discussion that specifies the QA reporting from the subcontractor to the prime contractor, along with the mechanism and frequency of such reporting. Regular QC reports from subcontractors during the course of a project are important in keeping project and QA management informed on progress and on potential problems which may require corrective action.

15.2 RI Field Quality Assurance Reports To Management

15.2.1 Frequency

At regular intervals (preferably at monthly intervals), the PMO QA manager will submit a quality assurance report to the program manager and the operations manager describing the performance of the quality assurance program for each site assignment. Problems or issues which arise between regular reporting periods may be identified to program management at any time.

15.2.2 Contents

The monthly quality assurance reports will contain:

- Results of system and performance audits conducted during the period
- An assessment of the measurement data, including accuracy, precision, completeness, representativeness, and comparability

- A listing of the nonconformance reports, including stop-work orders issued during the period, related corrective actions undertaken, and an assessment of the results of these actions
- Identification of significant quality assurance problems and recommended solutions

15.3 Laboratory Quality Assurance Reports To Management

This reporting system is a valuable tool for measuring the overall effectiveness of the QA program. It serves as an instrument for evaluating the program design, identifying problems and trends, and planning for future needs. Divisional QA directors submit extensive monthly reports to the vice president of QA and the divisional director. These reports include the following items.

- The results of the monthly system audits, including any corrective actions taken
- Performance evaluation scores and commentaries
- Results of site visits and audits by regulatory agencies and clients
- Problems encountered and corrective actions taken
- Holding time violations
- Comments and recommendations

In addition, on a monthly basis, a summary of the 5% QA audit of reported data is sent to the corporate QA office.

SECTION 16 REFERENCES

16.1 Generic References

References, if any, can be included in the body of the text, as footnotes, or collected in this section. If a reference is not readily available, attach a copy to the QA Project Plan. References must uniquely identify the cited material. In particular, when citing various compendia of standard methods published by the EPA, ASTM, American Public Health Association, etc., be sure to include the edition number, since such methods can change substantially from one edition to the next.

For example:

1. American Society for Testing and Materials. Annual Book of Standards. ASTM. Philadelphia, PA, Parts 14 and 19 (updated yearly).
2. U.S. Environmental Protection Agency. "A Procedure for Estimating Monofilled Solid Waste Leachate Composition." Technical Resource Document SW-924, 2nd ed. Hazardous Waste Engineering Research Laboratory, Office of Research and Development. Cincinnati, OH, and Office of Solid Waste and Emergency Response. Washington, D.C., 1986.
3. American Nuclear Society. ANSI/ANS-16.1-1988. American National Standard Measurement of the Leachability of Solidified Low-Level Radioactive Wastes by a Short-Term Test Procedure. American Nuclear Society. LaGrange Park, IL, 1986.

4. U.S. Environmental Protection Agency. Office of Solid Waste. Test Methods for Evaluating Solid Waste, 3rd ed. Available from U.S. Government Printing Office. Washington, D.C., 1986.
5. Alford-Stevens, A., T. A. Bellar, J. W. Eichelberger, and W. L. Budde. Method 680, Determination of Pesticides and PCB in Water and Soil/Sediment by Gas Chromatography/Mass Spectrometry. Available from the U.S. Environmental Protection Agency. Cincinnati, OH, 1985.
6. Klute, A. (Ed.). Methods of Soil Analysis, Part I. American Society of Agronomy. Madison, WI, 1986.

INDEX

A

AA, See Atomic absorption
Accuracy, 133–134, 136, 138, 155, 169, 172
Acetate buffers, 70–71
Acetic acid buffer, 67, 70–71
Acetone, 140
Acetone-hexane mixture, 68
Acid, GC sensitivity and, 88
Acid digestion, 35–37, See also Hot acid digestion
ACIL, 12
Additions, standard, 45–46
Aerosol carrier flow, 48–49
Air samples
  for metals analysis, 33–34
  for organic analysis, 68
Alkali metals, 59–60
ALM, 12
Alpha spectroscopy, 152
American Council of Independent Laboratories (ACIL), 12
American Public Health Association, 148
American Society for Testing and Materials (ASTM), 148, 177
Ammonium nitrate, NaCl removal method, 63–64
Analog to digital converter, 109–110
Analytical blanks, 139–140
Analyzer, for mass spectrometer, 105–107
Aqueous samples
  hot acid digestion, 35
  leaching extracts, 32
  preparation for organic analysis, 67–68
  sample size, 34
Argon
  ICP torch flow, 48, 57
  make-up gas mixture, 92, 97
Arochlor, 100–101
Arsenic, 59
ASCII format, 17
Association of Laboratory Managers (ALM), 12
ASTM, 148, 177
Atomic absorption (AA), 59–65, See also Graphite furnace atomic absorption
  calibration, 60
  interferences, 61–65
  matrix effect compensation (standard additions), 45–46
  matrix modification, 63–64
  maximizing sensitivity of, 60–61
  mercury determination, 44
  method QA requirements, 151
  sample digestion for, 35–36
  Zeeman background correction, 64–65
Atomic emission spectra, See Inductively coupled plasma
Audits, 162–164
Autosamplers, sample carry-over and, 89–90
Average percent recovery, 172

B

Background absorption, 46
Background correction, 64–65
Background subtraction, 58, 113–114
Backlog management, 4–8
  new tests, 10
  overbooking, 7
Baseline shifts, 55
Bases, GC sensitivity and, 88
Batch, defined, 43
Bench spike, 53
Benzene, 27–28

Blanks, 43–44, 139–140
  field equipment, 160
  quality assurance checklist, 24
Bombs, 33, 40–42
Bonded phase columns, 85
4-Bromofluorobenzene (BFB), 150
2-Butanone, 140

C

CAA, 24
Cadmium, 27
Calibration
  AA, 60
  GC/MS, 116
  ICP, 51–52
  nonlinear curve, 51
  QA project plan, 147–152
  radiological methods, 152
  reporting, 15
  standard additions and, 45
  temperature variability and, 52
Capacity, 4–5
  overbooking, 7
  reporting, 8
Capillary columns, 85
Carbonates, microwave digestion, 39
Carrier gases
  for GC, 90–93
    chromatographic resolution and, 91
    detector and, 92
    flow control, 92
  for ICP, 48
CERCLA, 24
Certification, 132
Chain of custody, 24, 142, 145–146, See also Sample custody
Characteristic wastes, 26
Chemical interferences,, 62
Chloroform, 151
Chromium, 27
Clean Water Act, 26
Client representatives, 25
Client services, 2–3
CLP, See Contract Laboratory Program
Columns, 85–88, 104
  mass spectrometer, 104, 108
  overloading, 116
Comparability, 134–135, 138
Competitiveness, 9
Completeness, 133–134, 136, 138, 169–170, 172
Computer systems, 17, 18
CONOSTAN, 43
Contamination, 44, 89–90
Contract Laboratory Program (CLP), 8, 23, 115
  reporting formats, 18–19
  shipping procedures, 145–146
Contract-required quantitation limits, 136
Control limits, 138
Copper ratio, 58
Corporate QA audit, 164
Corrective action plan, 173–175
Corrosive waste, 26
Cost
  equipment, 10
  new tests, 9
  staff, 11
Critical measurements, 128
Cross-contamination, 44, 89–90
Cryofocusing, 86
Cryogenic traps, 68
Custody of samples, See Sample custody
Customer base, 6–7
Customer management, 8
Cyanide, 27

D

Data base, 16, 17
Data management, 3, See also Information management
Data quality audits (DQAs), 162, 164
Data quality indicators, 168–173, See also Accuracy; Completeness; Precision
  field assessment, 170–171
  laboratory calculation, 171–173
Data quality objectives (DQOs), 21, 25, 132–140
  accuracy, 133–134, 136, 138
  comparability, 134–135, 138
  completeness, 133–134, 136, 138
  field quality control, 159
  method detection limit, 133–134, 136
  precision, 133–134, 136, 138
  representativeness, 135, 138
Data requirements, 135
Data transport, 17–18
Data use statement, 130
Data validation, 153–156
DB-1, 85
DB-5, 85

Dbase, 17
Debris, 32
Decafluorotriphenylphosphine (DFTPP), 110–111, 151
Detection limit, 25, 115, 172–173
  bomb-digested samples and, 42
  EPA definition, 101
  GC, 100
  GC/MS, 115–116
  mercury analysis, 44
  method detection limit, 134, 170
  mixed-phase samples and, 38
  nonstandard or modified methods, 150
  PCBs, 75
  purge and trap methods and, 70
  QA objectives, 133–134, 136
  sample size and, 34–35
Detectors, 93–101
  electrolytic conductivity, 98
  electron capture, 73, 92, 97–98
  flame ionization (FID), 69, 93–96
  GC carrier gas and, 92
  MS, 107–108
  photoionization, 98
DFTPP, 110–111, 151
1,1-Dichloroethene, 151
1,2-Dichloropropane, 151
Dilution
  AA interference compensation, 61–62
  ICP interference compensation, 53
  sample viscosity and, 42–43
  volatile-containing samples, 69
Dissolution methods, 42–43
Documentation and records, See also Reporting
  chain of custody, 24, 141, 145–146, See also Sample custody
  data reduction and validation, 156
  data reporting, 157
  field records, 143, 159
  instrument maintenance logbooks, 167
  nonconformance reports, 174
  personnel, 13
  project files, 157
  sample analysis, 147
  sample disposal, 147
  sample traffic report, 144
  sampling procedures, 141, 143–144
  stop-work order, 174
Double injections, 84
Down-time scheduling, 6
DQO, See Data quality objectives
Drinking water evaluations, 26
Duplicate control samples, 137–138, 161

E

Electrolytic conductivity detector, 98
Electron capture detector, 97–98
  carrier gas and, 92
  methanol extracts and, 73
Electron energy, 110
Employee records, 13
Environmental Compliance Agreements, 24
Environmental Protection Agency (EPA)
  CLP, See Contract Laboratory Program
  detection limit definition, 101
Equipment blanks, 160
Equipment cost, 10
Equipment preventive maintenance, 164–168
Ethylbenzene, 151
Evaporation, sample loss from, 37
Experimental design statement, 128
Extraction techniques, 67–68, 73–75

F

False negative results
  GC, 99–100
  GC/MS, 113–114
False positive results
  GC, 99–100
  GC/MS, 113–114
Ferrule leaks, 82
Field corrective action, 174
Field data
  assessment, 170–171
  reduction, 154
  reporting, 155
  validation, 154–155
Field preventive maintenance, 165–167
Field quality assurance audits, 162–164
Field quality assurance reports to management, 176–177
Field quality control checks, 159–160
Field records, 143, 159
Field Sampling Plan, 125
Filters
  acid wash, 35
  for air samples, 33
Flame atomic absorption, 59–60

Flame ionization detector (FID), 69, 93–96
  methanol extracts and, 73
  sample screening, 69
Floculants, 32
Flow control, GC carrier gas, 92
Flow rate
  atomic absorption spectrometry, 61
  GC/MS sensitivity and, 102–103
  ICP gas streams, 48–50
Foaming samples, 71–73
Fortified blanks, 43–44
Furnace atomic absorption, 59, See also Graphite furnace atomic absorption
  sample digestion for, 36

G

Gamma spectroscopy, 152
Gas chromatography (GC), 76–101
  acid/base behavior and, 88
  autosamplers and sample carry-over, 89–90
  columns, 85–88
  component temperatures, 88
  compound identification, 98–101
  detectors, 93–101, See also Flame ionization detector; Electron capture detector
    electrolytic conductivity, 98
    photoionization, 98
  false positives and negatives, 99–100
  film thicknesses, 85–86
  gases, 90–93
    chromatographic resolution and, 91
    detector and, 92
    flow control, 92
  headspace sampling system, 73–74
  injection speed, 89
  injector liners, 79
  injectors, 76–79, 82–84
  leaks, 81–82, 84
  methanol extracts and, 73
  method QA requirements, 150–151
  peak splitting, 84
  sample screening, 69
Gas chromatography-mass spectrometry (GC/MS), 101–117, See also Mass spectrometry
  avoiding reanalysis, 116–117
  column overloading, 116
  compound identification, 113–115
  data system, 17–18
  detection limits, 115–116
  low response problems, 102–105
  methanol extracts and, 73
  method QA requirements, 150–151
  scan speed and, 108
  target compound library, 113–115, 117
  tuning, 150
Gas proportional counters, 152
GC, See Gas chromatography
GC/MS, See Gas chromatography-mass spectrometry
GFAA, See Graphite furnace atomic absorption
Glassware
  capillary columns, 85
  heating vessel covering, 37
  mercury contamination, 44
Glow plug, 95
Graphite furnace atomic absorption (GFAA), 59
  analytical spike, 152
  interferences, 62
  matrix modification, 63–64
  sample digestion for, 36
  steps, 62
  temperature cycle, 63
  Zeeman background correction, 64–65
Graphite tubes, 60, 62
Groundwater data quality objectives, 136–137, See also Water samples

H

Halogens, flame ionization detector and, 93
Hardcopy data, 15, 25
Hazardous waste
  mixed-phase samples, 38
  treatment, 26–28
Health and Safety Plan, 125
Heating methods, 36
  bombs, 40–42
  glassware, 37
  hot plate, 36–38
  microwave, 38–40
Helium, 79, 92
HEPA filter, 33
Heterogenous samples, 32, 68–69
HETP, 91
Hexamethyldisilazane (HMDS), 79
Hexane
  acetone mixture, 27, 68
  oil dilution, 69
  PCB analysis in oils, 76
High-efficiency particulate air (HEPA) filter, 33
HMDS, 79
Holding time, 24

mercury determinations, 44
metals analysis, 35
report, 15
Hot acid digestion, 34, 35–37
heating methods, 36–42
Hot plates, 36–38
Hydrochloric acid-nitric acid mixture, 34, 35
Hydrogen gas, 91, 94
Hydrogen peroxide, 37–38
Hydroxides, 32

I

IAETL, 12
ICP, See Inductively coupled plasma
Identification, sample, 144
IFB procedures, 148
Igniter, 95
Inductively coupled plasma (ICP)
background subtraction, 58
calibration, 51–52
detection limits, 34
interferences
check samples, 56–57
matrix effects, 52–54
spectral interferences, 52, 54–56
internal standards, 43, 53–54
matrix effect compensation (standard additions), 45–46
method QA requirements, 151
MS-coupled, 34–36
nebulization for, 46–48
sample digestion for, 35–36
sequential and simultaneous instruments, 50–51, 58–59
torch and flow rates, 48–50, 57–58
Information management, 13–18
information categories, 14
laboratory information management systems, 17–18
manual tracking systems, 15–16
semiautomated systems, 16
Injection speed, 89
Injector, 76–79
foreign material in, 82–84
liners, 79
Inorganic analysis, 31–65, See also Metals analysis
Instrumentation
cost of, 10
data export capacity, 17
maintenance logbooks, 167
new tests and, 10
preventive maintenance, 164–168
training programs, 12
Instrument detection limit (IDL), 172
Interference, See Matrix effects; Spectral interference; Viscosity effects; specific analytical methods
Interference check samples, 56–57
Internal quality control checks, 157–161
Internal standard
GC/MS, 102–103
ICP, 43, 53–54
International Association of Environmental Testing Laboratories (IAETL), 12
Ionization energy, 110
Ion multiplier, 105, 107–108
Ion source, 104, 105
temperature, 110–111
Iron, vanadium baseline and, 55
Isomers, 102

J

Jet separator, 103

K

Kerosene, 42

L

Labels, 144
Laboratory capacity, 4–5
reporting, 8
Laboratory control sample (LCS), 44
Laboratory information management systems (LIMS), 17–18
Laboratory management, See Management
Laboratory organization, 1–4
Laboratory performance and system audits, 162–164
Laboratory preventive maintenance, 167–168
Laboratory quality assurance objectives, 137–144
Laboratory quality control checks, 160–161
Laboratory trade organizations, 12
Leachate samples, 67–68
data quality objectives, 137
Leaching, 32
Lead, 27, 32
copper ratio test, 58
gasoline determination, 33

Library, target compounds for GC–MS, 113–115, 117
Liquid-liquid extraction, 75
Listed hazardous wastes, 27–28
Lotus 123, 17
L'vov platform, 36, 60

M

MACRO, 17
Mail slot system, 15–16
Maintenance, 164–168
Make-up gas, 92, 97
Management, 1
  capacity and backlog, 4–8
  of clients, 8
  of data, 3, 13–18, See also Information management
  of personnel, 11–13
  of samples, 3–4, See also Sample custody
Manganese, copper ratio test, 58
Mass assignment drift, 106
Mass balance, 170
Mass spectrometry (MS), 101–117, See also Gas chromatography-mass spectrometry
  analog to digital converter, 109–110
  analyzer, 105–107
  calibration, 116
  columns, 104, 108
  compound identification, 113–115
  detector, 107–108
  instrument zero adjustment, 111
  internal standard, 102–103
  ionization energy, 110
  ion source problems, 105, 110–111
  leakage, 104–105
  pressure in, 103
  scan speed, 108–110
  target compound library, 113–115, 117
  tuning, 111–113
  types of, 102
Matrices, 31–34, 67–69
Matrix effects, 45
  AA, 62
  ICP, 52–54
  standard additions and, 45–46
  surrogate compounds and, 139
Matrix modification, for GFAA, 63–64
Matrix-specific control samples, 139
Mercury, 27, 44
Metal hydroxide sludges, 32
Metals analysis, 31–65, See also, See also Atomic absorption; Inductively coupled plasma
  field data assessment, 171
  holding time for, 35
  internal standards for, 43
  laboratory organization, 1–2
  laboratory QC samples, 161
  matrices, 31–34
  method QA requirements, 151–152
  mixed-phase samples, 38
  quality control samples, 43–44
  sample preparation, 34–45
    heating methods, 36–42
Methanol, 69, 73
Method blanks, 139–140
Method detection limit (MDL), 134, 170, 172–173
Methylene chloride, 27, 68, 139–140
  ultraviolet extraction, 74
Microwave heating, 33, 38–40
  organic sample preparation, 33
Mixed-phase samples, 38, 68–69
Mixed waste samples, turn-around times, 21
Monochrometer setting, 60
MS, See Mass spectrometry
Multiple ion detection (MID), 109, 111
Multiplier, 105, 107–108

N

National Pollutant Discharge Elimination System (NPDES), 24, 26, 32
Nebulization, 46–48
New Jersey, contract laboratory reporting formats, 18
Nickel, interference effects, 55–56, 64
NIOSH, 148
Nitric acid, 35–37
  HCl mixture, 34, 35, 45
  microwave-assisted digestions, 38
  silver digestion, 45
Nitrogen, GC/MS peaks, 104
Nonconformance reports, 174
NPDES, 24, 26, 32
NPDWA, 24

O

Occupational Safety and Health Administration (OSHA) requirements, 13

Oil samples, 33, 68
detection limits for inorganic analyses, 35
direct injection, 69
internal standards for, 43
microwave-assisted digestion, 39
PCB determination, 76
sample dissolution, 42
Organic analysis, See also Gas chromatography; Gas chromatography-mass spectrometry
dissolution methods, 42–43
field data assessment, 170–171
laboratory QC samples, 161
method QA requirements, 150–151
PCB determination, 76
sample preparation, 67–76
heterogenous materials, 68–69
matrices, 67–68
for PCBs and pesticide determination, 75–76
semivolatiles, 74–75
volatiles, 69–74
Organic samples
detection limits for inorganic analyses, 35
for metals analysis, 33
microwave-assisted digestion, 39
peroxide reactions, 38
Organization, 1–4
Organometallic compounds, 33
internal standards, 43
OSHA requirements, 13
OV-1, 85
OV-17, 85
OV-275, 85
Overbooking, 7
Oxygen, GC/MS peaks, 104

P

PAHs, See Polynuclear aromatic hydrocarbons
Part B permit, 28
PCBs, 75–76
Peak splitting, 84
Percent completeness, 138
Percent recovery, 169, 172
Performance and system audits, 162–164
Performance Evaluation Audits (PEAs), 162, 163
Permanganate digestion, 44
Peroxide, 37–38
Personnel, 11–13
flexibility, 7
responsibilities statement, 131–132, 153
training, 11–13
Pesticides, 75–76
PFTBA, 105, 111–112
pH
acetate buffers, 71
metal hydroxides and, 32
precision objectives, 134
toxic waste treatment, 26
Photoionization detector, 98
Phthalate esters, 140
Polyethyleneglycol, 69
Polynuclear aromatic hydrocarbons (PAHs), 75, 81
GC injection speed and, 89
photoionization detector and, 98
Practical quantitation limit (PQL), 100, 172–173
Precipitates, 35–36
ICP nebulizer and, 47
Precision, 133–134, 136, 138, 155, 171–172
calculation of, 168–169
Preparation blanks, 139–140
Pressure regulator, 92
Pricing, 9
Productivity, training and, 12
Project files, 157
Project planning, See Quality assurance project plan

Q

Quality assurance (QA), 3, 20–26
audits, 162–16
information management, 14
objectives, 132–140, See also Data quality objectives
officer (QAO), 162
reporting formats, 18
Quality assurance and control (QA–QC) data, 18
Quality Assurance Program Plan (QAPP), 125
Quality assurance project plan, 3, 20–21, 125
calibration and analytical procedures, 147
checklists, 21–24
corrective action plan, 173–175
data quality indicators, 168–173
data reduction, 152–157
field, 154–155
laboratory, 156–157
internal quality control checks, 157–161
performance and system audits, 162–164
preventive maintenance, 164–168
project description, 127–131
project organization and responsibilities, 131–132

quality assurance objectives, 132–140, See also Data quality objectives
references, 177–178
reports to management, 176–177
sample custody, 141–147
sampling procedures, 140–141
table of contents, 127
title and signature page, 125–126
Quality control checks, 157–161
Quality control samples, 43–44, 139
Quantitation report, 119–124

R

Radioactive samples, 21, 34
Radiological methods, 152
RCRA, 24, 26, 38
Reactive waste, 27
Reagent blanks, 139–140
Record keeping, See Documentation and records
References, for QA project plan, 177–176
Reflux, heating vessel covering and, 37
Regulatory drivers, 24
Relative percent difference (RPD), 133, 155, 168, 171
Relative standard deviation (RSD), 155, 168–169
Replication, 158
Reporting, 3, 8, 15, 18–20, 24, See also Documentation and records
field data assessment, 171
laboratory capacity, 8
laboratory data, 157
new tests, 10
nonconformance, 174
QA reports to management, 176–177
QC data, 153–154, 155
quantitation report (example), 119–124
Representativeness, 135, 138
Resource Conservation and Recovery Act (RCRA), 24, 26, 38
Responsibilities statement, 131–132, 153
Retention time, 98
Ribbed watch glasses, 37
Risk estimation models, detection limits for, 115–116

S

Sample custody, 141–147
collection documentation, 143–144
custody transfer and shipment, 145–146
field records, 143
laboratory procedures, 146–147
sample traffic reports, 144
Sample disposal, 25, 26–28
documentation of, 147
Sample dissolution, 42–43
Sample identification, 14, 144
Sample management, 3–4
Sample preparation, See Acid digestion; Air samples; Aqueous samples; Organic samples; Soil samples; Volatile components; Water samples; specific analytical methods
Sample receiving section, 11
Sample shipment, 145–146
Sample size
for bomb digestion, 42
detection limit and, 34–35
mercury analysis methods, 44
Sample storage, 146–147
Sample tracking, 13
manual systems, 15–16
semiautomated systems, 16
Sample traffic report, 144
Sample transport interferences, 52–54, 62, See also Matrix effects
Sample, definition, 4
Sampling and Analysis Plan (SAP), 125
Sampling containers, 25
Sampling logbook, 143–144
Sampling procedures, See also Aqueous samples; Air samples; Organic samples; Volatile components
QA project plan, 140–141
documentation, 143–144
quality control checks, 160
Scan speed, for MS, 108–110
Scandium, 43, 54
Scope of Work, 150
Security, 142, 146–147
Sediment samples, 31–32
data quality objectives, 137
Selenium, 59, 64
Semivolatile samples, 74–75
Sensitivity, See also Detection limits
flame ionization detector, 93–94
GC/MS, 102–105
scan speed and, 108

mercury analysis, 44
Sequential instrumentation, 50–51, 58–59
Serial dilution, for ICP calibration, 52
Shipment, 145–146
Signature page, for QA Project Plan, 125–126
Silicates, 34, 35
Silver, 27, 45
Simultaneous instrumentation, 50–51, 58–59
Single control samples, 137–138
Site description, 128, 129–130
Site manager, 175
Sludges, 32, 68, 73
Sodium chloride, GFAA analysis and, 63–64
Sodium sulfate, 74–75
Soil samples
data quality objectives, 137
detection limits, 34
extraction, 68
for metals analysis, 32
PCB determination, 75
potential data uses, 130
quality control checks, 160
sample size for, 34
volatile components, 73
Soil Sampling Quality Assurance User's Guide, 160
Solid samples, 68–69, 73–75, See also Soil samples
hot acid leaching, 37–39
Solvents, 33, 68–69
metals analysis of, 33
organic sample dissolution, 42–43
ultraviolet extraction, 74
waste treatment, 27
Solvent tailing, 79
Soxhelet extraction, 75
Spare parts inventory, 168
Spectral interference
AA, 60
ICP, 52, 54–56
interference check samples, 56–57
Spiked blanks, 43–44
Split valve, GC injector, 76–79
Spreadsheet, 16, 17
Staff, 7, 11–13, See also Personnel
Standard additions, 45–46
Standard deviation, 168–169
State-certified laboratory, 24
Stop-work order, 174
Storage, 25, 146–147
Subcontracts, 24
QA reports to management, 176
Sulfides, 27
Sulfuric acid, PCB analysis in oils, 76
Surface water, data quality objectives, 137
Surrogate, 137, 139
standards, 161
SW-846 methods, 38, 148–149, 160
Systems audits, 162–164

T

Table of contents, QA Project Plan, 127
Target compound library, 113–115, 117
Technical Systems Audits (TSAs), 162, 163
Temperature accuracy, 134
Temperature drift, ICP calibration and, 52
Tetraethylead, 33
Tetraglyme, 69
Thallium, 59
Tier II data, 19
Tier III data, 20
Tier IV data, 21
TMCS, 79
Toluene, 42, 117, 151
Total dissolved metals, 35
Toxic wastes, 26
Training, 11–13
Trimethylchlorosilane (TMCS), 79
Tuning, MS instrumentation, 111–113, 150
Turn-around time, 21, 23

U

Ultrasonic extraction, 74–75
Ultrasonic nebulizer, 48
Ultraviolet light, photoionization detectors, 98
Used instrumentation, 10

V

Validation, 153, 154–155
of nonstandard methods, 149–150
Vanadium, 55
Vinyl chloride, 151
Viscosity effects, 45
AA interferences, 62

dilution ratio and, 42–43
ICP interferences, 52
inorganic analysis and, 42
standard additions and, 45–46
Volatile components, 69–74
direct injection, 69
GC columns and, 86
headspace analysis, 73–74
purge and trap, 70–73

W

Waste analysis plan, 26–27
Waste treatment, 26–28, 32
Watch glasses, 37
Water samples, 31–32, 67–68
data quality objectives, 136–137
field quality control checks, 160
microwave-assisted digestion, 38
PCB determination, 75
potential data uses, 130
Wavelength setting, 50–51, 58
Wet chemistry tests, reagent blanks for, 140
Word Perfect, 17
Work plans, 23, 125

Y

Yttrium, 43, 54

Z

Zeeman background correction, 64–65